2012-2022 **中国天文学会成立100周年纪念**

CHINESE ASTRONOMY ON THE MARCH

崔向群◎主编

中国天文学会◎编

2022

 南京大学出版社

图书在版编目（CIP）数据

中国天文学在前进：2012—2022：汉、英 / 崔向群主编；中国天文学会编．—南京：南京大学出版社，2023.2
ISBN 978-7-305-26224-1

Ⅰ．①中… Ⅱ．①崔… ②中… Ⅲ．①天文学—文集—汉、英 Ⅳ．①P1-53

中国版本图书馆 CIP 数据核字 (2022) 第 208409 号

出版发行　南京大学出版社
社　　址　南京市汉口路 22 号　　邮　　编　210093
出 版 人　金鑫荣

书　　名　中国天文学在前进 (2012—2022)
主　　编　崔向群
编　　者　中国天文学会
责任编辑　王南雁　　编辑热线　025-83595840

照　　排　南京开卷文化传媒有限公司
印　　刷　苏州工业园区美柯乐制版印务有限责任公司
开　　本　787×960　1/12　印张 23　字数 325 千
版　　次　2023 年 2 月第 1 版　2023 年 2 月第 1 次印刷
ISBN　978-7-305-26224-1
定　　价　100.00 元

网　　址：http://www.njupco.com
官方微博：http://weibo.com/njupco
微信服务号：njupress
销售咨询热线：（025）83594756

中国天文学在前进

中国天文学会成立一百周年纪念文集

2012—2022

中国天文学会
中国 · 南京
2022

CHINESE ASTRONOMY ON THE MARCH

In Commemoration of the Centennial Anniversary of the Foundation of Chinese Astronomical Society

2012—2022

Chinese Astronomical Society

China · Nanjing

2022

Centennial Solemnization of
Chinese Astronomical Society (CAS)

中国现代天文学
和
中国天文学会的创始人和领导人

Founders and Leaders of
Modern Astronomy in China and CAS

高　鲁 | （1877–1947），福建人、曾八次任中国天文学会会长（9 年）

Gao Lu (1877–1947), Native of Fujian, Chairman of CAS for 8 times (9 years)

余青松 | （1897–1978），福建人、曾四次任中国天文学会会长（4 年）

Yu Qingsong (1897–1978), Native of Fujian, Chairman of CAS for 4 times (4 years)

张钰哲 | （1902–1986），福建人、院士、曾七次任中国天文学会理事长（39 年）

Zhang Yuzhe (1902–1986), Native of Fujian, Academician,

Chairman of CAS for 7 times (39 years)

王绶琯 | （1923–2021），福建人、院士、中国天文学会第五届理事长（1985–1989）

第六到十四届名誉理事长（1989–2021）

Wang Shouguan (1923–2021), Native of Fujian, Academician, 5th Chairman of CAS (1985–1989)

6th -14th Honorary Chairman (1989–2021)

李　珩 | （1898–1989），四川人、中国天文学会第一、二、三届副理事长（1957–1982）、第四届名誉理事长（1982–1985）

Li Heng（1898–1989）, Native of Sichuan, 1st, 2nd, and 3rd Vice-Chairman of CAS（1957–1982）, 4th Honorary Chairman （1982–1985）

陈遵妫 | （1901–1991），福建人、中国天文学会理事长（1946–1947）、第四届名誉理事长（1982–1985）

Chen Zungui（1901–1991）, Native of Fujian, Chairman of CAS（1946–1947）, 4th Honorary Chairman（1982–1985）

程茂兰 | （1905–1978），河北人、中国天文学会第二、三届副理事长（1962–1978）

Cheng Maolan (1905–1978), Native of Hebei, 2nd and 3rd Vice-Chairman of CAS (1962–1978)

戴文赛 | （1911–1979），福建人、中国天文学会第一、二、三届副理事长（1957–1979）

Dai Wensai (1911–1979), Native of Fujian, 1st, 2nd, and 3rd Vice-Chairman of CAS(1957–1979)

编辑委员会

撰稿人及资料提供者（按目录顺序）

景益鹏　唐正宏　张首刚　李孝辉　季海生　季江徽　王晓锋　李　成　宫雪非　史生才
吴雪峰　王祥玉　顾为民　熊少林　方　军　周礼勇　孙小淳　王广超　李向东　曹　晨
朱　进　余　恒　侯金良　崔辰州　李珊珊　田　斌　王　娜　宁宗军　汪景琇　陈鹏飞
田　晖　高　亮　黄光力　钱声帮　卢仙文　宋　谊　谌　悦　曾艺蓉　贺姝祎　黄鸾鸾
黄春萍　姚洁等

英文译者（按目录顺序）

景益鹏　唐正宏　张首刚　李孝辉　季海生　季江徽　王晓锋　李　成　宫雪非　史生才
吴雪峰　王祥玉　顾为民　熊少林　方　军　周礼勇　孙小淳　王广超　李向东　朱　进
余　恒　侯金良　崔辰州　安　涛　王　娜　宁宗军

提供大事记和图片的单位（按时间顺序）

国家天文台　云南天文台　新疆天文台　国家授时中心　紫金山天文台　上海天文台　南京天文光学技术研究所　北京天文馆　射电天文实验室　南京大学　北京大学　北京师范大学　上海交通大学　清华大学　中山大学　厦门大学　中国科技大学　江苏天文学会　福建天文学会　江西天文学会　山东天文学会　名词审定委员会　青年工作委员会　信息工作委员会　女天文工作者委员会　光电技术研究所等

英文审校

黄光力

UNITS PROVIDING CHRONICLE OF MAJOR EVENTS AND PICTURES
(ACCORDING TO TIME)

National Astronomical Observatories, Yunnan Astronomical Observatory, Xinjiang Astronomical Observatory, National Time Service Center, Purple Mountain Observatory, Shanghai Astronomical Observatory, Nanjing Institute of Astronomical Optics and Technology, Beijing Planetarium, Key Lab of Radio Astronomy, Nanjing University, Peking University, Beijing Normal University, Shanghai Jiao Tong University, Tsinghua University, Sun Yat-Sen University, Xiamen University, University of Science and Technology of China, Astronomical Society of Jiangsu, Fujian, Jiangxi, and Shandong, Working Committees of Astronomical Terminology, Young Astronomers, Astronomical Information, Female Astronomers, and Institute of Optics and Electronics, etc.

ADVISOR FOR THE TRANSLATIONS

Huang Guangli

目录

中国天文学砥砺前行的十年 景益鹏 __001

天体测量 唐正宏 __006
时间频率研究与应用 张首刚，李孝辉 __008
太阳物理 季海生 __011
行星科学 季江徽等 __013
恒星物理 王晓锋 __015
星系和宇宙 李成 __017
天文技术与方法 宫雪非等 __020
射电天文 史生才等 __022
空间天文和高能天体物理 吴雪峰等 __024
天体力学与卫星动力学 周礼勇等 __027

天文学史 孙小淳，王广超 __030
天文教育 李向东 __032
天文普及 曹晨，朱进 __035
天文学名词审定 余恒 __037
天文期刊 侯金良等 __039
天文图书 侯金良等 __041
天文信息 崔辰州，李珊珊 __043
青年天文工作者 田斌，安涛 __045
女天文工作者 王娜 __047
天文学会 宁宗军 __049

天文大事记（2012–2022）　__052
2012–2022 年获得国家一、二等奖和特等奖的天文项目（含国家级个人奖项）　__077
2012–2022 年获得省部级一等奖和特等奖的天文项目（含省部级个人奖项）　__078

中国天文机构　__081

图片部分　__195
一、 学会活动　__196
二、 院士风采　__203
三、 重大项目与成果　__210
四、 重要天象的观测　__228
五、 学术活动　__231
六、 天文教育　__246
七、 天文普及　__250

编后语　__259

CONTENTS

A Decade of Chinese Astronomy to Ahead Jing Yipeng __089
Astrometry Tang Zhenghong __097
Time-Frequency Study and Application Zhang Shougang, Li Xiaohui __100
Solar Physics Ji Haisheng __105
Planetary Science Ji Jianghui, et al. __108
Stellar Physics Wang Xiaofeng __112
Galaxies and Cosmology Li Cheng __115
Astronomical Technology and Methods Gong Xuefei, et al. __119
Radio Astronomy Shi Shengcai, et al. __122
Space Astronomy and High-energy Astrophysics Wu Xuefeng, et al. __124
Celestial Mechanics and Astrodynamics Zhou Liyong, et al. __128
History of Astronomy Sun Xiaochun, Wang Guangchao __133
Astronomical Education Li Xiangdong __136
Astronomical Popularization Cao Chen, Zhu Jin __140
Astronomical Terminology Yu Heng __142
Astronomical Journal Hou Jinliang, et al. __144
Astronomical Library Hou Jinliang, et al. __146
Astronomical Information Cui Chenzhou, Li Shanshan __148
Young Astronomer Tian Bin, An Tao __151
Female Astronomers Wang Na __153
Astronomical Society Ning Zongjun __155

A Chronicle of Major Astronomical Events (2012—2022) 159
Astronomical Achievements that Received National First, Second, and Special Prize in 2012–2022 (Including Personal Prize) 189
Astronomical Achievements that Received Ministerial/Provincial First and Special Prize in 2012–2022 (Including Personal Prize) 190

PICTURE 195
Activities of Chinese Astronomical Society 196
Academicians 203
Major Achievements and Projects 210
Observations of Important Celestial Events 228
Academic Activities 231
Astronomical Education 246
Astronomical Popularization 250

Word of the Editors 260

中国天文学砥砺前行的十年

中国天文学会第十四届理事长　景益鹏

党的十八大以来，我国天文学者在天文学各领域砥砺前行，在科研队伍的建设、天文设备的研制、科学论文的发表和引用等方面都取得了显著的成绩，在某些方面的指标已经位居世界前列。2021年恰逢中国共产党成立一百周年纪念，今年又是党的二十大即将召开之年，正当我国进入全面建设社会主义现代化国家并向着第二个百年征程进军的重要时刻，中国天文学会迎来了百年华诞。

在过去的十年里，中国天文学会在2014年和2018年召开了全国会员代表大会，共召开常务理事会、理事会50多次，共同研究和制定学会工作及方案。经过第十二届、十三届和十四届理事会的不懈努力，中国天文学会自身建设取得重大发展。会员单位增加了十个，他们分别是厦门大学天文学系、上海交通大学天文系、中山大学物理与天文学院、中国科学院光电技术研究所、苏州市青少年天文观测站、华中科技大学物理学院、上海科技馆、云南大学、西华师范大学、山东大学空间科学与物理学院，平均每年发展个人会员50余人，与山东天文学会、重庆天文学会建立业务指导关系。目前，中国天文学会有单位会员33个，个人会员3300余人，与17个省市天文学会和2个省市天文小组建立业务指导关系。在原有的组织、教育、普及、名词审定、图书信息出版工作委员会的基础上，顺应形势的发展，在2016年至2019年间，先后成立青年工作委员会、女天文工作者委员会和信息工作委员会。至此，中国天文学会由8个工作委员会、11个专业委员会和基金工作组构成。

通过天文学会各专委会和工委会下属成员的共同努力，在过去十年里，我国天文学各分支均取得令人瞩目的成果，在一系列重大天文设备的研制和科学问题的研究方面均取得了重要的进展。在宇宙大尺度结构的理论模型和计算机模拟、银河系大规模光谱巡天、太阳结构和爆发机制、深空探测和太阳系天体等方面优势明显。随着LAMOST望远镜的投入使用，和500米口径球面射电望远镜（FAST）、高海拔宇宙线观测站（LHAASO），及空间天文项目“悟空”号暗物质粒子探测卫星、“慧眼”号硬X射线调制望远镜、“羲和”号太阳双超卫星等重大科学基础设施项目的陆续建成，重大科研成果产出持续增加，共有11篇研究论文在*Nature*和*Science*刊物上发表，为我国迈向天文强国奠定了坚实的基础。十年来共获得包括发现高能电子宇宙射线能谱超出、大样本恒星演化与特殊恒星的形成、基于高精度脉泽天体测量的银河系旋臂结构研究的三项国家自然科学二等奖、54项国家和省部级集体及个人奖项。天文学者入选时代楷模和“五四”集体奖，说明天文学者在国内的影响力大幅提高；还有女天文学家获得全国先进工作者和三次中国女青年科学家奖，说明女天文学者的地位明显上升。

过去的十年，我国天体测量学在基础研究和应用领域都取得多项重要成果。系统分析研究了太阳系绕银河系运动引起的长期光行差效应对河外参考架的影响，从超过40年的VLBI观测数据中精确解算出长期光行差的数值，为第三代国际天球参考架（ICRF3）的建立做出显著贡献。同时，自主编制了一部银盘外天区的绝对自行星表，该星表覆盖了银盘外（银纬正负30度以外）2.2万平方度的天区，计算给出了暗至R波段亮度为20.8星等的1亿多颗天体的位置和绝对自行数据，且包含从光学到近红外7个波段的测光数据。联合研制一台高精度底片扫描仪，完成全国3万张时间跨度近百年的天文历史底片的数字化，相关结果已纳入中国虚拟天文台。

过去的十年，是时间频率研究蓬勃发展的十年，随着北斗卫星导航系统、国家重大科技基础设施——高精度地基授时系统、空间站高精度时间频率实验系统等国家重大项目的部署和实施，我国在时间频率领域飞速发展，自主创新能力显著增强，取得了一系列可喜的成果，世界时不再依赖国外数据。国家授时中心在2019年建成了融合数字天顶筒、VLBI2010、iGMAS等系统的地球自转测量与数据服务系统，并为“嫦娥五号”、“天问一号”等重大航天计划提供世界时数据。依托洛南40米望远镜，我国建成了脉冲星计时观测系统，并利用7年15颗毫秒脉冲星相关观测数据建立综合脉冲星时间尺度，开展脉冲星时驾驭原子钟试验研究。经过几十年的攻关，国产小铯钟研发终于取得了系列突破，成功研发了原子态光抽运和光检测的光抽运小铯钟，并形成了批产能力，装备应用于北斗导航、海军导航、5G通信、电力运行、轨道交通及北极科考等系统，保障了国家标准时间产生发播等重要基础设施运行的自主可控。同时，我国自主研发了主要性能与美国产品同水平、体积仅为其一半的8立方厘米小芯片原子钟产品，并成功装备应用在微自主导航定位授时终端、北斗导航特殊接收终端、水下导航系统和某通信系统等。另外，自行研制被动型氢原子钟也在北斗导航卫星成功运行，长短波授时系统亮相新中国成立70周年成就展和中国共产党党史馆。

近十年中国已经成为国际太阳物理界的重要研究力量。在科研方面，我国每年发表在国际期刊上的文章大约250篇，仅次于美国。中科院出具的2014–2018年天文学领域文献计量统计报告显示，在全球1%和10%高引用论文中，来自中国的太阳物理论文占全球的比例均比中国天文学的平均水平要高2～3倍。影响力指数（某国特定学科的篇均被引频次与该学科全球篇均被引频次的比值）为0.997，表明中国的太阳物理论文在引用上已达到国际平均水平。我国学者通过分析国内外观测数据或模拟，在太阳大气的基本辐射和动力学过程、太阳大气的精细结构的形成以及相关的波和能量的传输、太阳爆发活动的储能与释能物理机制、太阳磁场长期演化及发电机的模拟方面获得了大量的进展，这些进展增加了人类对于发生在太阳上的辐射磁流体力学（或磁化等离子体）等基本物理过程的认识，为灾害性空间天气的准确预报提供了新的认识。在设备方面，“羲和”号太阳卫星已经发射，先进天基太阳天文台即将发射，将进一步提升我国的国际地位。

我国天文学家在行星地球动力学、太阳系小天体观测和研究、系外行星系统研究、天体化学研究等方面取得了重要成果，5次入选年度中国十大天文科技进展。首次发现了地球自转参数之日长变化中存在显著的约8.6年周期的振幅增强信号，揭示该振荡的极值时刻与地磁场快速变化的发生存在密切的对应关系；新发现27个近地小行星，包括5个对地球构成潜在威胁的小行星，建立小天体光变数据库，揭示主带族群热物理特性；精确确定图塔蒂斯小行星轨道，使“嫦娥二号”实现千米级飞越探测，基于“嫦娥二号”探测数据，揭示图塔蒂斯的物理特性、自转特性、内部结构与形成机理及表面地质特征与演化历史；基于LAMOST光谱数据和系外行星与其宿主恒星的刻画，发现新的热海星族群；利用南极的CSTAR和AST–3望远镜发现上百颗系外行星候选体；利用“嫦娥五号”样品开展矿物岩石学研究，发现一种相对罕见的月海高钛玄武岩，为深入了解月球晚期岩浆活动和内部月幔源区物质地球化学特性提供精确的证据。

随着大视场光谱巡天望远镜LAMOST、硬X

射线调制望远镜 HXMT、500 米口径球形射电望远镜 FAST 等一批国家重大科研装备的建成和成功运行，我国在恒星及银河系领域的研究实现了快速发展，实现了观测手段从高能到射电的全波段覆盖，整体研究水平与国际先进水平的快速接近。在此期间，我国恒星物理领域研究人员发表 2000 余篇文章。利用 LAMOST 巡天获得了世界上最大的传世恒星光谱库；基于“慧眼”的观测，获得了中子星、黑洞和伽马射线暴的大量观测数据，发现了最高能量的中子星基频回旋吸收线、距离黑洞最近的相对论喷流和快速射电暴的第一个 X 射线对应体，并且证认其来自于银河系内的一颗磁陀星。利用 FAST 发现了大量的脉冲星，并使得我国在快速射电暴及中性气体的研究位居国际前列；发现 4 例高色散新快速射电暴，重复暴的偏振时变，严格限制首例河内暴源的流量，迄今最大的快速射电暴爆发事件样本等。

我国天文学家在宇宙学理论模型、暗物质探测实验和宇宙学观测限制等方面都取得重要成果。理论方面，开展了宇宙起源模型（特别是反弹模型和暴胀模型）、多种暗能量模型、修改引力模型、原初黑洞和引力波的产生机制、宇宙早期相变、中微子宇宙学等理论研究。实验方面，建成了锦屏山地下实验室，启动基于液氙的 Panda-X 和基于高纯锗的 CDEX 两项暗物质直接探测实验，获得国际先进的暗物质模型参数限制；我国天文学家完成了一系列具有国际领先地位的数值模拟，其中纯暗物质模拟包括“盘古”、“九天”、ELUCID、Cosmic Growth、Cosmic Zoom、“凤凰”和 TianNu；基于目前的 X 射线观测，首次指出银河系大尺度上热气体的存在；基于“悟空”号卫星数据精确测量宇宙线正负电子、质子和氦核能谱，发现能谱新结构，获得对伽马射线线谱辐射最灵敏的搜寻结果。LHAASO 发现首批“PeV 加速器”和最高能量光子，测定高能天文标准烛光蟹状星云的超高能伽马射线亮度，开启了“超高能伽马天文学”新时代。

我国在天文技术与方法研究领域自主成功研制出一批面向不同观测对象、覆盖不同观测波段的望远镜和科学终端，例如 LAMOST 和 FAST 的终端、中国空间站工程巡天望远镜（CSST、口径 2 米）的巡天模块等；1.8 米太阳望远镜 CLST 工程实现初光并在调试中；发展了国际前沿的太赫兹天线与探测终端技术，包括 1.4THz 频段最高灵敏度超导热电子混频器和我国首台太赫兹超导相机，兴隆 2.16 米望远镜、丽江 2.4 米望远镜配置了高分辨率光纤光谱仪。这些望远镜和科学仪器的研制和使用不但为我国天文学长远发展奠定了基础，同时提升了我国自主研制大型望远镜和终端设备的创新能力。

“中国天眼”500 米口径球面射电望远镜（FAST）于 2016 年 9 月落成，2020 年 1 月通过国家验收并正式向国内开放。在“早出成果、多出成果、出好成果、出大成果”的要求下，取得了一批重大科研成果：通过原创的中性氢窄线自吸收方法，FAST在测量星际磁场方面取得了重大进展；FAST在新兴的快速射电暴研究方向产生了一系列重要发现，包括4例高色散新快速射电暴，重复暴的偏振时变，严格限制首例河内暴源的流量，迄今最大的快速射电暴爆发事件样本等；FAST共发现约500颗脉冲星等。新建成的天马65米射电望远镜大大增强了CVN对深空探测器实时精密测定轨能力，出色完成了探月工程“嫦娥二号”、“嫦娥三号”、“嫦娥四号”、“嫦娥五号”和火星探测器“天问一号”的相关任务；在国家需求应用层面，由佘山25米、乌鲁木齐26米、密云50米和昆明40米射电望远镜组成的中国VLBI网（CVN）在我国深空探测工程中发挥了不可替代的作用。

在天文教育、天文学史、天文期刊、名词审定、信息化以及青年天文工作者和女天文工作者等方面，过去的十年我国都取得长足的进步。2012年到2021年是中国天文教育发展最迅速的十年，目前共有34所高校开展天文相关的科学研究，22所成立了天文（学）系开展天文学本科教育，其中12所是最近十年成立。天文科普参与全国科技活动周、全国科普日等活动。于2020年举办了“仰望星空”科学之夜和科技活动周重大示范活动——“一带一路”科普交流周，于2021年5月参与举办了“中国天文学会空间天文探测科普”学术交流活动。

展望未来，中国天文学会正在向下一个一百年征程迈进。结合国家发展战略，中国天文学将与国家发展同行。随着科学技术的不断进步，我国天文工作者将不断开拓创新，将更多的新技术、新方法、新设备持续不断地使用到天文学研究中。空间卫星先进天基太阳天文台即将发射，地面设备包括国家重大科研仪器研制项目“2.5米大视场高分辨率太阳望远镜”（WeHosT）已经正式立项，冷湖2.5米大视场巡天望远镜项目（WFST）、2米环形太阳望远镜、慕士塔格1.9米光学望远镜、丽江1.6米多通道测光巡天望远镜（Mephisto）正在研制之中。另外，中国大型光学红外望远镜、6.5米宽视场巡天望远镜（MUST）、LAMOST二期工程、中国巨型太阳望远镜（CGST）、司天工程、110米全可动射电望远镜、15米亚毫米波望远镜等重大项目正在积极推进之中。这些项目的建成将有力推动我国天文学的发展，使中国天文学早日进入国际发达水平前列。

天体测量

唐正宏

近 10 年来，我国天体测量学在基础研究和应用领域都取得多项重要成果。

在天文参考系方面，系统分析研究了太阳系绕银河系运动引起的长期光行差效应对河外参考架的影响，从超过 40 年的 VLBI 观测数据中精确解算出长期光行差的数值，为第三代国际天球参考架（ICRF3）的建立做出显著贡献。利用新一代测地 VLBI 观测数据，基于相位闭合约束条件，解算出部分河外射电源结构及其变化。借助最新的 Gaia 资料，全面分析了 ICRF3 的特性，为 ICRF3 的有关参数提供了独立的验证和补充分析。统计分析了河外源的光学与射电位置的偏差，给出了这种偏差与河外源射电结构（尤其是喷流结构）相关的直接证据，为光学和射电参考架连接提供了第一手参考资料。厘清了银道坐标系转换和应用过程中由于光行差椭圆项而可能存在的误解，指出 IAU1958 定义的随着参考系转换而带来的问题。首次提出用现代大型巡天星表，在较长波段的星表来直接构建最符合银河系结构的坐标系，并讨论了基于银道面和银心观测资料建立正交坐标系的方法，结果表明采用 IAU1958 坐标系会产生大约 0.5° 的指向差异；该成果有可能在将来推动由中国学者提出的新 IAU 决议的形成。在广义相对论框架下，摒弃了黄道的概念，采用最新的太阳系历表，得到一组高精度长期岁差参数。

在地球自转方面，创新性地把基于人造卫星激光测距数据解算得到的地球形状因子 J2 应用到地球岁差模型的解算中来，在积分岁差方程时考虑了地球形状因子 J2 的减速效应，以及天球和地球参考架的自洽性，在将来 IAU 讨论新的标准岁差模型时，可提供重要的参考。

在高精度天体测量数据处理方面，受哈勃空间望远镜高精度天体测量定标的启发，提出了 CCD 图像几何扭曲精确求解的自定标方法。已应用于国内主要的几台大口径望远镜的高精度定标中，明显地提高了（包括天然卫星，近地小行星等天体在内的）天体测量精度。此外，还提出了利用 Gaia 星表的色指数精确求解较差颜色折射的天体测量改正方法、多 CCD 芯片之间间距的精确求解方法。这些方法不仅可以应用到地面望远镜的高精度天体测量中，还可以推广应用到我国即将发射的空间站巡天望远镜（CSST）的高精度天体测量定标中。在地面望远镜天体测量中，还成功揭示出两靠近天体位置测量精度提高的特征范围，这对于太阳系小天体的观测将起到指导的作用。此外，还提出了小行星相邻两晚稀疏位置观测资料精确测量目标地心距离的方法，新方法将对新发现的小行星、近地小行星等的观测具有良好的应用价值。

系统开展了巨行星（木星、土星、天王星和海王星）天然卫星的CCD观测，获得大量CCD图像资料，解算出一批天然卫星的高精度天体测量位置，被美国JPL用于编制DE440/ DE441历表。利用自主开发的轨道计算程序，对海卫一、海卫二等天然卫星轨道进行了改进。与英法学者合作，开展了Cassini探测器观测图像的归算和资料处理工作，取得了良好的进展。

利用独创的同时性较差观测的原理，基于Gaia实测的数据解算卫星基本角的变化，与其他监测结果交叉比对，从而核实并独立获得基本角变化，为Gaia最终的科学数据的修正提供重要的基准。

自主编制了一部银盘外天区的绝对自行星表，该星表覆盖了银盘外（银纬正负30度以外）2.2万平方度的天区，计算给出了暗至R波段亮度为20.8星等的1亿多颗天体的位置和绝对自行数据，且包含从光学到近红外7个波段的测光数据。该工作入选中国十大天文科技进展。

实现了我国探月工程中嫦娥二期三号任务——月球紫外望远镜（LUT）的天体测量支持系统（其中包括导星跟踪和观测策略）的方案制定和实施等工作，2012年12月“嫦娥三号”登月后，基于此套软件月球紫外望远镜顺利实现了第一架可无人调试和值守的全自动望远镜，获取了大量十分宝贵的观测资料。

在科技部科教基础性专项重大项目支持下，联合研制一台高精度底片扫描仪，完成全国3万张时间跨度近百年的天文历史底片的数字化，相关结果已纳入中国虚拟天文台。在上海市科委支持下，研发了2台气浮导轨扫描仪，其中一台具备彩色和反射扫描功能；在科技部的支持下，完成对意大利都灵天文台和乌兹别克斯坦兀鲁伯天文研究所近2万张天文历史底片的扫描。

时间频率研究与应用

张首刚，李孝辉

过去的十年，是时间频率研究蓬勃发展的十年，随着北斗卫星导航系统、国家重大科技基础设施——高精度地基授时系统、空间站高精度时间频率实验系统等国家重大项目的部署和实施，我国在时间频率领域飞速发展，自主创新能力显著增强，取得了一系列可喜的成果。

世界时不再依赖国外数据。世界时描述地球的空间方位，在空间科学、国防等领域必不可少。自20世纪90年代起，传统的世界时测量手段退役后，我国的世界时完全依靠国外数据，存在安全隐患。中科院国家授时中心和上海天文台、国家测绘局等单位开展了相关研究。国家授时中心在2019年建成了融合数字天顶筒、VLBI2010、iGMAS等系统的地球自转测量与数据服务系统，并为“嫦娥五号”、“天问一号”等重大航天计划提供世界时数据。

脉冲星时研究取得突破进展。新疆25米、昆明40米、上海65米和洛南40米等射电望远镜均具备脉冲星计时观测能力，贵州500米射电望远镜已新发现近300颗脉冲星，并开展着毫秒脉冲星计时观测研究。国家授时中心依托洛南40米望远镜，建成了脉冲星计时观测系统，并利用7年15颗毫秒脉冲星相关观测数据建立综合脉冲星时间尺度，开展脉冲星时驾驭原子钟试验研究。中科院高能所和航天科技集团开展了空间X射线脉冲星导航观测试验研究。

国产小铯钟打破发达国家垄断。受限于复杂的科学和工艺难题，在2017年之前的半个世纪内，国际上守时小铯钟一直被美国磁选态小铯钟产品独家垄断。2018年以来美国对我国完全禁售。经过几十年的攻关，国产小铯钟研发终于取得了系列突破，兰州空间技术物理研究所完成了磁选态小铯钟产品；北京大学联合湖北汉光公司等完成了磁选态光检测小铯钟产品；国家授时中心联合成都天奥电子公司和中国电科十二所，研发了原子态光抽运和光检测的光抽运小铯钟，并形成了批产能力。光抽运小铯钟主要性能超过了美国产品，装备应用于北斗导航、海军导航、5G通信、电力运行、轨道交通及北极科考等系统，保障了国家标准时间产生发播等重要基础设施运行的自主可控，并被国际权度局推荐用于各国标准时间的产生。多台产品对国际标准时间产生做出了贡献，取得了权重。

自研世界最小体积原子钟打破国外垄断。具有芯片大小、重量和功耗的微型原子钟，是基于量子物理理论和微机电加工集成技术，实现的较高性能的时间频率信号产生器件，作为部件为相关设备运行提供标准时间频率参考信号。十几年来，美国一直独家垄断国际市场。近几年来，美国开始限制芯片原子钟出口中国。中科院国家授时中心和精密测量院组织院内相关力量，自主研发了主要性能与美国产品同水平，体积仅为其一半的 8 立方厘米小芯片原子钟产品，并成功装备应用在微自主导航定位授时终端、北斗导航特殊接收终端、水下导航系统和某通信系统等。团队还实现了主要性能相同，体积 2.3 立方厘米的原理样机。研究成果入选国家“十三五”科技创新成绩展。

光学频率原子钟研究进入国际并跑行列。基于其优异性能，光学频率原子钟将取代铯原子微波频率钟，作为基准钟重新定义时间单位“秒”，具有前瞻性。中国计量科学研究院、华东师范大学、华中科技大学、中科院精密测量研究院和国家授时中心等单位陆续开展着不同元素的光学频率原子钟研制，均取得了不同进展，进入了国际先进行列。

世界首台冷原子钟在空运行。基于地面冷原子喷泉钟原理和空间站微重力环境，中科院上海光机所研制的匀速直线飞行的冷原子束微波钟，在“天宫二号”飞行器上成功运行。包括空间站时间频率产生系统、空地时间频率比对系统和地面测试及应用系统的我国空间站高精度时频实验系统转入正样研制阶段，近期将在空间运行世界首台冷原子光学频率原子钟。上海天文台和航天科工集团研制被动型氢原子钟在北斗导航卫星成功运行。北斗导航系统性能国际先进，并开始向全球提供导航定位及授时服务。

长短波授时系统亮相新中国成立 70 周年成就展和中国共产党党史馆。通过基准钟研制应用、守时钟研制应用、异地联合守时、守时算法改进、国际比对优化等手段，我国国家标准时间 UTC（NTSC）与国际标准时间偏差从 10ns 提升到 5ns 以内，对国际标准时间的贡献权重从世界第五提升到世界第三，为我国长波、短波，及北斗、长河等授时系统提供了可靠、稳定的时间溯源服务。我国已经基本形成了包括电话、网络、光纤、无线电短波、无线电长波，以及卫星等多手段的授时服务体系，较好地支撑了经济社会运行和国家安全发展。

时间频率测量比对基本仪器设备实现技术突破。时间间隔测量设备的测量范围和精度，频率稳定度测量分析仪的本地噪声性能达到国际领先，但作为常规测量仪器，其通用性和环境适应性需要进一步改进提升。国家授时中心通过与企业合作在开展着相关产品研发工作；光学频率测量是时间频率测量研究和发展的热点和趋势。国家授时中心研制的飞秒光学频率测量仪器主要性能国际先进，应用在不同的精密测量领域。应用自主研发的超稳定度频率激光器和飞秒激光频率梳，实现了超稳光生微波频率信号源，较常温频率信号源频率稳定度提高两个量级。其应用把国家授时中心冷原子喷泉钟短期频率稳定度提高了半个量级。卫星共视、卫星双向比对、长波定时，以及远程时间比对与复现等仪器设备实现了国产化，并装备应用在重大设施系统。基于北斗卫星实现的欧亚时间比对技术，被国际组织推荐用于国际时间比对。

高精度时间频率传递技术快速发展。光纤是目前可用的最高精度时间频率远程传递技术手段。应用通信运营商光纤，实现千公里级光纤时间频率高精度传递，包括光学频率信号、微波频率信号和时间信号。量子时间同步具有精度高和安全性强的特点，国家授时中心应用 50km 光纤实现了亚皮秒的时间同步稳定度，为远程量子时间同步奠定了基础。基于光纤时间频率传递技术和增强型罗兰授时技术，国家授时中心承担了国家重大科技基础设施“高精度地基授时系统”研制建设任务。

社会的不断发展和进步，对时间频率的技术和服务提出了越来越高的要求。未来，随着空间站时间频率实验系统，高精度地基授时系统等重大设施的建设和运行，我国将实现更高性能、更具弹性、更加融合的国家时间基准产生与服务体系，有力支撑科技强国建设，有力支撑经济社会高质量发展。

太阳物理

季海生

在过去十年里，随着我国太阳物理研究队伍的不断壮大，中国已经成为国际太阳物理界的重要研究力量。在科研方面，我国每年发表在国际期刊上的文章大约 250 篇，仅次于美国。中科院出具的 2014–2018 年天文学领域文献计量统计报告显示，在全球 1% 和 10% 高引用论文中，来自中国的太阳物理论文占全球的比例均比中国天文学的平均水平要高 2 ~ 3 倍。影响力指数（某国特定学科的篇均被引频次与该学科全球篇均被引频次的比值）为 0.997，表明中国的太阳物理论文在引用上已达到国际平均水平。我国学者通过分析国内外观测数据或模拟，在太阳大气的基本辐射和动力学过程、太阳大气的精细结构的形成以及相关的波和能量的传输、太阳爆发活动的储能与释能物理机制、太阳磁场长期演化及发电机的模拟方面获得了大量的进展，这些进展增加了人类对于发生在太阳上的辐射磁流体力学（或磁化等离子体）等基本物理过程的认识，为灾害性空间天气的准确预报提供了新的认识。值得一提的是，大多数工作是由年轻学者牵头完成。

在地面观测方面，我国目前正在常规运行的光学观测台站有怀柔基地、抚仙湖基地等，子午工程Ⅱ期磁场望远镜和 H-α 望远镜即将落户紫金山天文台赣榆站。自 2014 年起抚仙湖的“新真空太阳望远镜（NVST）”已经发表了一百多篇审稿论文。代表性成果包括对色球小尺度磁重联过程的直接成像观测，首次在太阳耀斑中发现具有扭缠结构磁岛形成的快速磁重联、黑子或者网络场旋转形成扭缠磁结构，并触发太阳活动和高清晰的日珥结构等。日冕观测取得了零的突破。我国研制成功多套毫秒级高分辨宽带动态频谱仪，装备于国家天文台、云南天文台和山东大学等站址。此外，国家专项也将建设超宽带高分辨频谱系统；山东大学研制成功 35 ~ 40GHz 的毫米波射电观测系统。基于我国宽带动态频谱仪在过去 20 年积累的数据，发现了一批独特的微波频谱精细结构和尖峰群及耀斑射电前兆，并发展了诊断爆发源区磁场和粒子加速过程的新方法。2016 年建成了明安图厘米 - 分米波射电日像仪 MUSER，具备了宽带频谱成像能力。国家子午工程Ⅱ期正在明安图和稻城各布局建设一套米 - 十米波日像仪，二者的互补观测完整覆盖 30 ~ 450 MHz 的频率范围。在过去十多年里开展了西部太阳选址工作，在川西无名山堪选了大口径地基太阳望远镜（包括日冕仪）的台址；此外，在青海冷湖堪选了红外太阳观测的台址。8 米级光学望远镜的科学目标业已完成，立项过程在推动之中；地基大视场 2.5 米望远镜和 2 米环形望远镜正在建设之中。

在空间观测方面，H-α 光谱望远镜（CHASE）成功发射已经取得科学数据，标志着我国太阳物理界的空间观测取得实质性突破，同时，先进天基太阳天文台（ASO-S）将在 2022 年择机发射，其科学应用系统基本建好。数个空间（预研）项目已被提出，主要聚焦在偏离日地连线方向的立体观测，其中包括在黄道面上布点的立体或全景观测，也包括飞离黄道面从太阳极区方向进行的立体观测；另外，我国学者提出的具有创新特色的项目，如紫外光谱仪观测日（星）冕磁场和立体磁场观测，也已经提出并得到充分预研。

在空间天气预报方面，国内多家单位成立了相应的研究队伍，经验预报模型和数值模拟预报模型都取得一定的效果。近十年来观测数据迅猛增加，人工智能技术越来越多地用来自动提取前兆特征，并取得良好的预报效果。学者们成功利用机器学习方法从紫外辐射预测了日冕软射线辐射。在 CME 的对地有效性预报方面，近十年获得长足进步。近期研究表明，机器学习方法对 CME 抵地时间的预报表现出色，误差缩短到6小时左右，与人工预报相似。研究也表明对太阳爆发进行立体观测将进一步提高预报精度。

在类太阳恒星活动研究方面，通过多年的努力，我国太阳物理学者已与国际主流保持同步。代表性工作有：（1）基于 Kepler 和 TESS 空间望远镜的恒星光变数据，建立了恒星磁场和耀斑的活动性指标，并与太阳活动进行了比较，构建了恒星耀斑活动数据集，分析了恒星耀斑活动的统计分布规律；（2）利用 LAMOST 望远镜的恒星光谱观测数据，构建了恒星色球活动光谱数据集；（3）获取了恒星色球活动的统计规律，分析了恒星色球活动性与恒星年龄的关系。以 LAMOST 望远镜为代表的自主大型巡天观测设备及其海量观测数据是我国在这一研究领域的优势所在。

本报告主要参考了汪景琇、陈鹏飞、田晖等人撰写的《太阳物理规划》（2020）。

行星科学

季江徽等

十年来，我国天文学家在行星地球动力学、太阳系小天体观测和研究、系外行星系统研究、天体化学研究等方面取得了重要进展。在 *Nature* 期刊发表 1 篇文章，在 *Nature Astronomy* 发表 3 篇文章，在 *Nature Communications* 发表 1 篇文章，在 *PNAS* 发表 3 篇文章，在高水平学术期刊（*ApJ* 系列杂志、*A&A* 和 *MNRAS*）上发表约 300 篇文章，研究成果 5 次入选年度中国十大天文科技进展。

在行星地球动力学方面：首次发现了地球自转参数之日长变化中存在显著的约 8.6 年周期的振幅增强信号，揭示该振荡的极值时刻与地磁场快速变化的发生存在密切的对应关系；融合 JUNO 与 Cassini 高精度重力场测量数据，建立了木星和土星的自洽内部结构模型与大气环流深部构造；提出木星稀释核结构的形成机制；通过局部谱分析方法，揭示了火星北半球岩石圈的强磁化区域埋藏深度较浅，而南半球岩石圈的强磁化区域埋藏深度深得多；系统性地分析了类地行星自转轴进动和热不稳定性共同驱动的液核流体动力学现象，发现局部化热对流条件。

在太阳系小天体观测和研究等方面：近地天体搜寻发现、物理性质与监测预警等方面取得系列成果，新发现 27 个近地小行星，包括 5 个对地球构成潜在威胁的小行星，建立小天体光变数据库，揭示主带族群热物理特性；精确确定图塔蒂斯小行星轨道，使“嫦娥二号”实现千米级飞越探测，基于“嫦娥二号”探测数据，揭示图塔蒂斯的物理特性、自转特性、内部结构与形成机理及表面地质特征与演化历史，分别入选 2012 年度和 2013 年度中国十大天文科技进展；利用三维彗核形状演化模型分析柯伊伯带小天体 Arrokoth 扁平结构的形成机制，提出了物质挥发导致太阳系小天体形状形成和演化的机制，揭示柯伊伯带原始星子的形成机理；获得了中红外月球的形貌特征及随月相变化规律，揭示了其物理本质，利用卫星开展了月食研究，揭示月食光度变化特征并发现了多个热点地区；原位月球太空风化研究揭示风化导致光谱斜率减小成因，发现波长小于 2.3 μm 存在热辐射，揭示微尺度月球热辐射特性。

在系外行星系统研究方面：融合 LAMOST 和 Kepler 数据揭示系外行星偏心率分布特征并约束

行星形成，入选2016年度中国十大天文科技进展；基于LAMOST光谱数据和系外行星与其宿主恒星的刻画，发现新的热海星族群，入选2018年度中国十大天文科技进展；利用南极的CSTAR和AST-3望远镜发现上百颗系外行星候选体，入选2019年度中国十大天文科技进展；系统研究系外行星形成演化和动力学，基于行星生长、行星盘和气态巨行星等揭示了近共振构型成因的新机制，发现大气逃逸导致行星尺寸在两个地球半径附近呈现分布低谷，定量刻画HL Tau原行星盘的精细结构特征，获得江苏省科学技术二等奖；基于ALMA观测和数值模拟研究揭示原行星盘中尘埃环结构成因、旋涡结构证据和旋涡产生的旋臂结构的特征；基于大样本热木星的光谱测光的透射谱观测，在其大气中发现以碱金属原子压力致宽线翼为表征的晴空特征，首次探测到锂的疑似信号，揭示钠原子、钾原子、水分子、氧化钛分子等大气组分；采用高分辨率多普勒光谱细致刻画经典热木星大气，提供不同类型行星的高层大气加热机制的观测依据。

在天体化学研究方面：利用首批申请到的两块“嫦娥五号”样品开展矿物岩石学研究，发现一种相对罕见的月海高钛玄武岩，为深入了解月球晚期岩浆活动和内部月幔源区物质地球化学特性提供精确的证据；在月球陨石中发现迄今为止太阳系中最大的氯同位素分馏；研究了火星陨石的岩矿学、微量元素地球化学和年代学特征，揭示火星表面发生的玄武岩岩浆喷发事件和复杂的冲击变质历史；研究了多种小行星（如灶神星）陨石的热变质和冲击变质历史，揭示小行星母体分异时间和机制；开展了玻璃陨石等高精度钾同位素分析，探究了中等挥发性元素钾在太阳系内的亏损和分馏机制。

我国天文学家积极开展了国际交流与合作：与美国加州大学圣克鲁兹分校开展了系外行星动力学研究；与德国马普太阳系研究所开展了彗发三维辐射转移模型研究等。参加近地天体国际联测，代表中国参与联合国框架下的IAWN国际组织。参与国际合作计划或观测设备对系外行星开展研究，如通过KMTNe（微引力透镜），GAIA（天体测量），SPIRou（视向速度），LCOGT（凌星）搜寻和观测行星，GTC探究行星大气，ALMA描绘原行星盘特征，SDSS刻画宿主恒星等。近十年，通过人才培养和引进，形成了一支结构合理的研究队伍，有6名学者入选国家级人才计划。我国天文学家正规划以系外宜居行星探测为核心科学目标的空间项目，如“近邻宜居行星巡天”计划，“地球2.0”计划与“紫瞳”计划等。

恒星物理

王晓锋

在过去的十年中，随着大视场光谱巡天望远镜 LAMOST、硬 X 射线调制望远镜 HXMT、500 米口径球形射电望远镜 FAST 等一批国家重大科研装备的建成和成功运行，我国在恒星及银河系领域的研究实现了快速发展，实现了观测手段从高能到射电的全波段覆盖，整体研究水平与国际先进水平的快速接近。在此期间，我国恒星物理领域研究人员在 *Science* 上发表论文 2 篇，在 *Nature* 上发表论文 8 篇，在 *Nature* 子刊上发表论文近 20 篇，在专业一流期刊（*ApJ* 系列杂志、*A&A* 和 *MNRAS*）上发表 2000 余篇文章。恒星物理领域的高被引用文章（>200 次）数在国内天文领域中占比较高，在总数 16 篇中占到其中的 8 篇（50%）。

利用 LAMOST 巡天获得了世界上最大的传世恒星光谱库。基于这一光谱数据，得到了大样本恒星年龄及金属丰度的分布规律，得到了一批特殊丰度恒星样本并限制其起源，精确描绘了银盘三维恒星质量分布及银盘不同位置的恒星形成历史，给出了不同年龄星族的银盘径向和垂向金属丰度梯度和分布函数，发现了银河系恒星盘可能具有进动的翘曲结构，发现了银河系内大质量恒星级黑洞候选体。基于 LAMOST 光谱及 Kepler 数据，利用星震学首次区分红巨星与红团簇星，发现白矮星内核存在氧元素超丰现象，发现超级耀斑恒星呈现增强的磁活动。同时结合其他巡天数据，发现了银河系恒星晕内扁外圆的结构和速度椭球各向异性，描绘出延伸到 100kpc 的银河系旋转曲线，给出了太阳邻域的暗物质密度的可靠估计。

“慧眼”硬 X 射线调制望远镜的建成运行弥补了我国在高能领域观测的短板，显著提升了我国在致密天体领域研究的影响力。基于“慧眼”的观测，获得了中子星、黑洞和伽马射线暴的大量观测数据，产生了一批重要成果：发现了最高能量的中子星基频回旋吸收线，发现了距离黑洞最近的相对论喷流，发现了快速射电暴的第一个 X 射线对应体，并且证认其来自于银河系内的一颗磁陀星，伽马射线暴探测率国际第二。FAST 是国际上单面口径最大的射电望远镜，利用该设备的观测发现了大量的脉冲星，并使得我国在快速射电暴及中性气体的研究位居国际前列。两台设备联合观测成功限制了快速射电暴的磁陀星起源的机制。

除了基于重大设备的研究之外，我国学者利用国内外设备开展了分子云 / 恒星形成、银河系磁场、银河系旋臂结构、超新星及宇宙学等方向的研究，并在国际上占有一席之地，具体成果包括：完成了银道面偏振巡天，基本厘清了邻近半个银盘的大尺度磁场结构，并首次揭示出银河系晕中的环向磁场结构；精确测量了太阳附近英仙座旋臂的距离，描绘了太阳附近几个 kpc 范围内的旋臂结构；对北银道面分子云开展了 CO 及其同位素三条谱线同时巡天，发现了新的银河系旋臂结构；利用多种设备的巡天数据，获取恒星及星际介质的精确消光规律；发现数百颗近邻星系的新爆发超新星，发现 Ia 型超新星存在两类前身星通道，进一步提高了其宇宙学测距精度。

在理论研究上，发展和完善了描述恒星对流的新模型，解决了一些长期存在的问题；精确确定了一批恒星的内部结构和演化状态，为恒星微观物理过程提供了新约束，在双星物质交换的稳定性判据研究上取得重要进展；在国际上开拓和推广了双星星族合成在星族、星系研究中的应用，利用该方法在特殊恒星（如 Ia 超新星、热亚矮星、X 射线双星等）的研究上取得了重要突破；提出了伽马射线暴的中心能源可能是处于超临界吸积状态的黑洞或高度磁化快速旋转的中子星（毫秒磁星），解释了暴后能源的持续活动性。

未来，随着一批地面及空间观测设备的建成和投入运行，如科大 - 紫台 2.5 米巡天望远镜、云大 1.6 米望远镜、中法 SVOM 卫星、爱因斯坦探针以及中国空间站的 POLAR-2 和 CSST 等，我国有望在恒星物理领域的重大科学问题的研究上取得进一步突破。

星系和宇宙

李成

过去十年来，我国天文学家利用国内外观测设备和数值模拟，在星系宇宙学各方面取得了大批研究成果，使我国在本领域的国际影响力持续增长。

我国天文学家在宇宙学理论模型、暗物质探测实验和宇宙学观测限制等方面都取得重要进展。理论方面，开展了宇宙起源模型（特别是反弹模型和暴胀模型）、多种暗能量模型、修改引力模型、原初黑洞和引力波的产生机制、宇宙早期相变、中微子宇宙学等理论研究。实验方面，建成了锦屏山地下实验室，启动基于液氙的 Panda-X 和基于高纯锗的 CDEX 两项暗物质直接探测实验，获得国际先进的暗物质模型参数限制；发射了“悟空”号卫星，测量具有高能量分辨率的宇宙线能谱并开展暗物质间接探测。观测限制方面，结合国内外多种天文观测数据（如 Planck、BOSS 和 eBOSS 等）和新兴的统计方法（如最优红移加权、多源天体交叉关联、切片 Alcock-Paczynski 方法、空洞统计、红移畸变的数值－理论混合建模、深度学习等），加强了对宇宙学模型的观测限制，尤其是在较高的置信水平发现暗能量的动力学迹象。建立了国内首个具有自主知识产权且处于国际领先水平的 FourierQuad 弱透镜测量管线，成功应用于 CFHTLS 和 DESI 等图像巡天。

我国天文学家完成了一系列具有国际领先地位的数值模拟。其中纯暗物质模拟包括“盘古”、“九天”、ELUCID、Cosmic Growth、Cosmic Zoom、“凤凰”和 TianNu，首次精确测量了暗晕速度偏袒，定量计算出轴子暗物质对宇宙结构形成的影响，重构了近邻宇宙的真实物质分布和结构演化历史，在跨越 21 个量级的空间尺度上精确解析了暗晕内部物质分布，在星系团尺度上对暗物质湮灭信号给出精确预言，利用宇宙原初密度扰动重构暗晕和星系角动量并首次发现星系角动量与宇宙原初密度扰动的相关性。流体模拟方面，“NIHAO”项目通过国际合作完成了 100 个星系的高分辨率流体模拟，兼具大样本和高分辨的双重优点，系统地研究了矮星系到银河系尺度的星系形成过程。此外，基于巡天统计，建立了解析的条件质量函数演化模型，刻画了暗晕中星系的质量演化和恒星形成历史；结合数值模拟、星系巡天和 Gaia 观测数据，给出了银河系质量的最精确测量。

我国天文学家实施了多项中高红移星系和类星体巡天观测。其中宽场多色图像巡天“南银冠U波段巡天（SCUSS）”和“北京－亚利桑那巡天（BASS）”分别被国际光谱巡天项目SDSS-IV/eBOSS和DESI采用作为选源数据。“再电离时期莱曼－阿尔法星系巡天（LAGER）”探测到一批红移7左右的莱曼发射线星系（LAE），首次揭示了再电离的非均匀性，发现宇宙最遥远的原初星系团。与国际团队合作实施的“CFHT大天区U波段深场巡天（CLAUDS）”以及自主实施的红移z=2.3的LAE和HAE窄带图像巡天，都为中高红移星系的探测和统计研究提供了第一手资料。此外，基于LAMOST等国内望远镜发现了大批红移5以上的极亮类星体和变脸类星体，发现了宇宙早期发光最亮、黑洞质量最大的类星体。

在低红移星系观测方面，我国天文学家全程深度参与了国际上最大规模的积分场光谱巡天SDSS-IV/MaNGA，领导了其中约三分之一的科学研究课题并取得了大批重要进展，包括：恒星初始质量函数的变化规律、恒星形成活动的熄灭规律、小质量和大质量星系的恒星形成历史、恒星尘埃消光的分布规律、星系各组分的动力学模型和暗物质比例、气体和恒星运动学的不一致现象、活动星系核的宿主特征、恒星形成与活动星系核和星系相互作用的关联等。同时，利用国内望远镜开展了一批近邻大星系的多缝光谱观测，结合国际积分场光谱数据开展了系列研究。

我国天文学家充分利用国内外观测设备，在星系气体、星系（周）际介质和活动星系核反馈等方面开展多波段观测研究。使用FAST望远镜和SKA探路望远镜ASKAP，首次对多个近邻星系团（群）获得连续的HI成图。基于星系气体的多波段观测提出了延展型恒星形成定律。利用Keck望远镜积分场光谱仪，在红移2发现一类巨型Lyα星云并首次发现星系周气体的面亮度随红移存在演化。基于目前的X射线观测，首次指出银河系大尺度上热气体的存在；在星系M81中心黑洞周围探测到高温高速外流风，证实了理论预言的热吸积流；探测到活动星系核kpc尺度的延展铁线发射及其与星系尺度的冷气体成协。利用类星体光变的新方法，首次针对活动星系核所驱动的高速气体外流开展大样本统计分析。

此外，我国天文学家还从中等质量黑洞和黑洞潮汐瓦解事件的搜寻、高红移黑洞－星系质量关系、黑洞吸积盘高速内流、宽吸收线类星体外流速度轮廓、解析宇宙X射线背景辐射等多个角度研究黑洞与星系的内在关系及其宇宙学意义。

展望未来，我国已建成运行FAST望远镜，正在研制阿里宇宙微波背景辐射极化望远镜（AliCPT）、空间站巡天望远镜（CSST）、LAMOST-II望远镜、中国科学技术大学－紫金山天文台2.5米大视场巡天望远镜（WFST）、清华大学宇宙热重子巡天望远镜（HUBS）和大视场多目标巡天望远镜（MUST）等观测设备，中国参与的国际合作项目DESI、PFS、SKA等也正在或即将付诸实施，我国的星系宇宙学研究领域必将在下一个十年取得更大的发展。

天文技术与方法

宫雪非等

近十年来，我国在天文技术与方法研究领域取得了很大进展，自主成功研制出一批面向不同观测对象、覆盖不同观测波段的望远镜和科学终端。这些望远镜和科学仪器的研制和使用不仅为我国天文学长远发展奠定了基础，同时提升了我国自主研制大型望远镜和终端设备的创新能力。

郭守敬望远镜（LAMOST）作为我国天文界的第一个国家重大科技基础设施于2012年开启正式光谱巡天，目前已观测了2000万余条光谱，成为世界上第一个获取光谱数超千万量级的光谱巡天项目。LAMOST发布的千万光谱数据和不断涌现的突破性研究成果，都证明了其前瞻性的设计理念和自主创新技术的成功。500米口径球面射电望远镜（FAST）通过国家验收并正式开放运行，这是目前世界上最大且最灵敏的单口径射电望远镜。FAST超高的灵敏度优势已经并将持续在脉冲星搜索、快速射电暴起源、星系形成及演化、引力波探测等领域产生具有深远影响力的科学成果。高海拔宇宙线观测站（LHAASO）进入科学运行阶段，这是目前世界上规模最大、灵敏度最高的超高能伽马射线巡天望远镜，和能量覆盖最宽广的宇宙线观测站。

中国空间站工程巡天望远镜（CSST）已正式立项，这台口径2米的空间望远镜配备了包括巡天模块在内的五台第一代仪器，将是中国天文科学迈向国际前沿的重大机遇。首颗用于空间天文观测的暗物质粒子探测卫星“悟空”（DAMPE）成功入轨并投入观测，获得了迄今为止世界上最精确的宇宙射线电子、质子、氦核等TeV以上能区的测量结果。“慧眼”硬X射线调制望远镜卫星是中国第一颗空间X射线天文卫星，使我国在高能天体物理观测领域占有了重要的一席之地。首颗太阳探测科学技术试验卫星“羲和号”（CHASE）成功发射，首颗综合性太阳探测卫星先进天基太阳天文台（ASO-S）将于2022年下半年发射升空，实现太阳磁场、太阳耀斑和日冕物质抛射的同时观测。面向时域天文学和高能天体物理的X射线天文卫星爱因斯坦探针（EP）正处于工程研制阶段。首台星载氢原子钟顺利升空，为北斗系统精密定位及自主导航性能提升奠定了技术基础。

国家重大科研仪器研制项目“2.5米大视场高分辨率太阳望远镜”（WeHosT）正式立项。冷湖2.5米大视场巡天望远镜项目（WFST）、2米环形太阳望远镜、慕士塔格1.9米光学望远镜、丽江1.6米多通道测光巡天望远镜（Mephisto）正在研制之中。1.8米太阳望远镜CLST实现初光并在调试中，获取太阳的高分辨成像的试观测图。1.2m望远镜激光测距系统首次实现地月间激光测距，填补了我国在月球激光测距领域的空白。无人值守南极巡天望远镜（AST3）在昆仑站投入观测，是我国唯一探测到引力波事件GW170817光学对应体的望远镜。光学和近红外太阳爆发监测望远镜（ONSET）和“新一代厘米–分米波射电日像仪”成功建成并投入科学观测。亚洲最大的上海65米射电望远镜投入运行，圆满完成“嫦娥五号”采样返回VLBI测轨定位及“天问一号”探测近火捕获任务。

十年来，在科学仪器研制方面也取得了长足进步。发展了国际前沿的太赫兹天线与探测终端技术，包括1.4THz频段最高灵敏度超导热电子混频器和我国首台太赫兹超导相机。兴隆2.16米望远镜、丽江2.4米望远镜配置了高分辨率光纤光谱仪。国际合作重大项目GTC望远镜超稳定高分辨率光谱仪将成为北半球最具竞争力的恒星视向速度测量装置，与TMT望远镜、MSE望远镜、Hale望远镜等国际重大项目共同研制的科学仪器也在稳步进展之中。

展望未来，中国大型光学红外望远镜、6.5米宽视场巡天望远镜（MUST）、LAMOST二期工程、中国巨型太阳望远镜（CGST）、司天工程、110米全可动射电望远镜、15米亚毫米波望远镜等重大项目正在积极推进之中，南极冰穹A和我国西部的优良台址也得到进一步确认。随着科学技术的不断进步，在我国天文工作者不断创新努力下，更多的新技术、新方法、新设备将会持续不断地投入到天文学研究中，有力推动我国天文学的发展。

射电天文

史生才等

在这十年里，我国射电天文事业在望远镜建造及科学研究方面取得了世界瞩目的成绩，在学科向广度和深度方向发展、人才队伍建设等方面也得到了快速的进步。

“中国天眼”500米口径球面射电望远镜（FAST）于2011年3月开工，2016年9月落成，2020年1月通过国家验收并正式向国内开放。在“早出成果、多出成果、出好成果、出大成果”的要求下，取得了一批重大科研成果：通过原创的中性氢窄线自吸收方法，FAST在测量星际磁场方面取得了重大进展；FAST在新兴的快速射电暴研究方向产生了一系列重要发现，包括4例高色散新快速射电暴，重复暴的偏振时变，严格限制首例河内暴源的流量，迄今最大的快速射电暴爆发事件样本等；FAST共发现约500颗脉冲星，成为自其运行以来国际上发现脉冲星效率最高的设备，而且在深入研究特殊脉冲星方面取得多项重要成果，如首次获得脉冲星三维速度与自转轴共线证据。此外，与费米伽马射线天文台大视场望远镜（Fermi-LAT）的天地一体化协同观测开启了研究脉冲星电磁辐射机制的新途径。

基于国内其他射电望远镜（天马65米、乌鲁木齐26米、德令哈13.7米等）及国际观测设备，我国天文学者在射电天文领域也取得了令人瞩目的成绩，例如：银心糖分子的成图观测，一氧化碳、氨分子、甲醛分子谱线巡天；星际脉泽研究，及基于脉泽高精度天体测量的银河系旋臂结构研究；恒星形成活动在星系尺度、极端贫金属星系、银河系内的表现，及磁场对其影响；银河系分子云的大尺度分布、样本检测、距离测量；超新星遗迹与分子云的相互作用；普朗克冷分子云团块的动力学和化学性质；亚毫米波瞬变源搜寻；脉冲星和磁星的周期跃变、测时噪声、辐射特征的研究；构建银河系电子密度模型YMW16；活动星系核的长期射电流量监测和短时标快速光变监测等。

在国家需求应用层面，由佘山25米、乌鲁木齐26米、密云50米和昆明40米射电望远镜组成的中国VLBI网（CVN）在我国深空探测工程中发挥了不可替代的作用；新建成的天马65米射电望远镜大大增强了CVN对深空探测器实时精密测定轨能力，出色完成了探月工程“嫦娥二号”、“嫦娥三号”、“嫦娥四号”、“嫦娥五号”和火星探测器“天问一号”的相关任务；发展的 VLBI 双波段 ΔDOR 及实时动态双目标同波束 VLBI 等技术也在我国探月工程中发挥了重要作用。

在望远镜建设及终端仪器研制方面，FAST 之外：天马 65 米口径射电望远镜作为我国自主研制的第一台大型全可动射电望远镜，综合性能居世界前列；奇台 110 米口径全可动射电望远镜（QTT）得到中科院和新疆维吾尔自治区共同支持，并获得立项批复；完成了南极 5 米太赫兹望远镜天线的方案设计和缩比模型验证，成功研制了国际前沿水平的超导接收机和超导阵列成像系统、无人值守的超宽带太赫兹傅立叶光谱仪等。

作为东亚天文台的主要成员之一，管理运行詹姆斯·克拉克·麦克斯韦尔望远镜（JCMT）。成为 SKA 政府间组织正式成员，并在我国建设了 SKA 区域中心原型机。2021 年 FAST 向全球开放。参与 VLBI、EVN 和 EHT 的协同观测。与多个国家建立了双边合作论坛机制，包括中美射电天文科学与技术学术研讨会、中澳天体物理论坛等。

空间天文和高能天体物理

吴雪峰、王祥玉、顾为民、熊少林、方军

近十年来，随着我国一批天基设备如“悟空”号暗物质粒子探测卫星(DAMPE)、天宫二号“天极”伽马暴偏振探测仪（POLAR）、“慧眼”号硬X射线调制望远镜（Insight-HXMT）、“极光计划”X射线偏振测量立方星（PolarLight）、“天格计划”空间分布式伽马暴探测网（GRID）、“怀柔一号”引力波暴高能电磁对应体全天监测器（GECAM）、“羲和”号太阳H-α光谱探测与双超平台科学技术试验卫星（CHASE）等相继发射和地基设备如南极巡天望远镜（AST3）、500米口径球面射电望远镜（FAST）、高海拔宇宙线观测站（LHAASO）等陆续建成并投入使用，我国学者在空间天文和高能天体物理领域取得了系列研究突破。

在引力波事件和伽马暴研究方面：南极AST3望远镜成功观测首例双中子星并合引力波事件GW170817的光学对应体，慧眼卫星对GRB170817A的MeV辐射性质提供了严格的上限约束；参加了基于欧南台甚大望远镜VLT观测项目，获得了国际上最完整的GW170817千新星光谱和偏振观测资料，并对该事件的伽马暴、千新星及多波段余辉开展了全面研究；从伽马暴的历史数据中发现了数例与千新星/并合新星理论模型预言一致的信号；发现从重子火球过渡到磁化火球演化的特殊伽马暴、史上最短的Ⅱ型伽马暴；统计发现伽马暴X射线余辉的自组织临界现象；POLAR获得国际最大的高精度伽马暴瞬时辐射偏振测量样本；从Chandra数据中发现首例双中子星并合形成的磁陀星所驱动的X射线暂现源。这些研究有力约束了中子星物态方程与极强磁场等基本物理。“怀柔一号”卫星发现一批伽马暴和磁星爆发，并突破在轨实时触发和下传观测警报的技术，引导多波段望远镜开展联合观测。

在快速射电暴（FRB）研究方面：利用慧眼卫星发现首个快速射电暴（FRB 200428）的X射线对应体并认证其来自于河内磁陀星（SGR J1935+2154），利用FAST观测给出磁陀星射电辐射流量的严格上限，发现X射线暴与FRB的弱相关性，入选*Nature/Science* 2020年国际十大科学进展/突破；利用FAST发现重复暴FRB 180301偏振角演化多样性，揭示了重复暴FRB 121102的能量分布呈双峰结构；提出中子星吸积小行星、双中子星绕转、中子星吸积白矮星、夸克星壳层坍缩等一系列原创性起源和辐射机制模型，率先开展 FRB 宇宙学和基本物理研究。

在脉冲星和中子星研究方面：FAST 开启了大规模的脉冲星发现和深度研究脉冲星进程，首次找到了脉冲星三维速度与自转轴共线的证据，并开展了高精度测时等工作，为未来引力检验等科学目标奠定了基础；利用国内外天文望远镜，在极低频引力波搜寻、脉冲星计时阵、脉冲星辐射等多个方向取得了系列研究进展；“极光计划”捕捉到了来自蟹状星云发生自转突变后 X 射线偏振信号的变化与恢复；在超新星遗迹 Kes 79 南侧发现 X 射线暂现低磁磁陀星；应邀完成星际和星系际磁场的 *ARA&A* 评述。

在 X 射线双星研究方面：首次测定了 X 射线极亮天体中黑洞的质量；首次在超软 X 射线源中发现相对论重子喷流；利用“慧眼”卫星直接测量到迄今宇宙最强磁场，发现距离黑洞最近的相对论喷流，发现高速逃离黑洞的高温等离子体；提出利用毫赫兹准周期振荡给出中子星半径下限的方法，并给出 4U1636–53 中子星半径约大于 11 公里。

在活动星系核（AGN）研究方面：利用氢、氦吸收线多普勒红移，首次观测到类星体中给吸积盘直接提供吸积燃料的快速内流；提出宽线区起源于尘埃环的理论，破解了宽线区起源和物理结构问题，提出了测量宇宙学距离的新方法；系统研究了 AGN 吸积流中存在的风，应邀为 *ARA&A* 撰写黑洞吸积理论的综述论文；采用贝叶斯方法对 1392 个 Fermi 耀变体进行分类并讨论了其辐射机制；在耀变体 PKS 2247–131 的伽马射线辐射中首次发现了月量级的准周期振荡现象，为喷流中的螺旋磁场模型提供了直接的观测证据；求解了强磁主导喷流径向平衡方程的解析解；参与黑洞事件视界望远镜（EHT）国际合作。

在高能宇宙线和伽马射线研究方面：基于“悟空”号卫星数据精确测量宇宙线正负电子、质子和氦核能谱，发现能谱新结构，获得对伽马射线线谱辐射最灵敏的搜寻结果；西藏羊八井 AS γ 实验发现超高能宇宙线加速候选天体，首次发现 PeV 能量宇宙线源存在于银河系的证据；LHAASO 发现首批“PeV 加速器”和最高能量光子，测定高能天文标准烛光蟹状星云的超高能伽马射线亮度，开启了“超高能伽马天文学”新时代；从 Fermi 数据中发现第一个有伽马射线辐射的超亮红外星系 Arp 220，将星系红外 – 伽马辐射的相关性延展至更高的光度范围；利用 Fermi–LAT 数据发现 M31 星系的伽马射线辐射集中在星系中心，而非沿星盘分布。

在高能中微子研究方面：提出活动星系核喷流的多辐射区模型，解释了跟耀变体关联的高能中微子的起源；提出超大质量黑洞的潮汐瓦解事件产生高能中微子源的模型，解释了与潮汐瓦解事件AT2019dsg关联的高能中微子的起源；根据IceCube对伽马暴中微子的观测，对伽马暴喷流中的质子含量进行了限制。

未来几年，我国研制的先进天基太阳天文台（ASO-S）、爱因斯坦探针（EP）、天基多波段空间变源监视器（SVOM）等卫星将按照计划在“十四五”期间发射运行，中国空间站工程巡天望远镜（CSST）已正式立项，我国已正式成为平方公里阵列射电望远镜（SKA）国际组织成员国并预计于2025年建成SKA-1。国内高校与科学院合作研制的多台两米级光学时域巡天望远镜将在“十四五”期间建成投入使用。我国空间天文与高能天体物理的研究将继续保持高速发展态势。

天体力学与卫星动力学

周礼勇等

近十年来，随着太阳系内及太阳系外行星系统观测快速进展、深空探测计划的逐步实施，天体力学和卫星动力学的研究对象、理论内涵都得到了丰富和扩展。

1. 天体力学基础理论

上海天文台和南京大学分别在摄动函数展开方面分别开展了工作。建立了除碰撞奇点外全局收敛的摄动函数展开方法，适用于包括交叉轨道的情形，可对近地小行星、彗星、类冥王星以及特洛伊等天体的轨道长期动力学演化进行理论分析。利用新的摄动函数展开处理方法，得到了高倾角乃至逆行平运动共振的相空间结构随轨道根数变化的情况。对非限制性三体问题轨道稳定性开展了系统性理论研究，给出了轨道翻转和轨道交换等特殊动力学现象的参数空间。建立了适用于任意自转空间定向的行星内部潮汐耗散模型，展现了其与第三体摄动共同作用导致行星轨道迁移的动力学过程。

2. 太阳系小天体动力学

南京大学及紫金山天文台等对柯伊伯带天体、特洛伊天体的运动展开了系统的研究。深入分析多个共振的高轨道倾角柯伊伯带天体的动力学，模拟重现了海王星轨道迁移并俘获此类天体的过程，预报了此类天体的发现位置，并对行星的迁移给出了更多限定条件。

对特洛伊天体动力学的系统研究彻底解决了当前太阳系构型下此类天体的轨道稳定性问题，完整刻画了影响各行星特洛伊的动力学机制。预报了海王星、天王星的特洛伊发现区域及几率；否定了地球、金星附近存在尚未发现的原初特洛伊天体的可能性。

通过统计所有小天体轨道角动量之和的方式，发现柯伊伯带矮行星的重要影响，并对可能存在的第九行星的质量和轨道参数给出了限定条件。

3. 系外行星系统

南京大学、紫金山天文台和上海天文台等将

传统天体力学研究拓展至系外行星系统领域。

对单恒星周围的行星系统：提出巨行星扰动小行星带物质使其落入白矮星的模型，建立了环白矮星尘埃盘和白矮星大气金属污染的统一演化图景；发现借助 Gaia 卫星的时序天体测量数据，可以反演一阶平运动共振的基本特征；行星动力学理论结果被应用于证认行星、统计行星系统轨道特性。

对环双星的行星系统：提出类地行星通过星子散射轨道向内迁移的机制；给出了“行星密近凌星”的发生条件，指出该现象导致的凌星时长增加有利于精确测量行星大气。

对星团环境下的行星系统：发现疏散星团中的多行星系统稳定性较好，大部分单恒星即使被散射出星团，也能保持其原有的行星系统；分析了星团中大质量恒星的光致蒸发效应对原行星盘寿命的影响；统计得到了行星出现率与运动速度的相关性，给出行星系统出现率与恒星动力学历史相关联的证据。

4. 天文动力学

上海天文台等单位深度参与北斗导航系统建设。承担了“北斗二号”、“北斗”试验支持系统、“北斗三号”信息处理系统研制任务，设计并实现了我国首套实时、多类型海量测量数据的信息处理系统。建立了基于北斗星地双向时间比对信息约束的 GEO 卫星精密定轨处理方法，突破 GEO 卫星高精度实时精密定轨难题；基于北斗星间链路数据，设计了星地星间联合精密定轨与时间同步方法，突破地面区域监测网对 MEO 卫星轨道和钟差高精度测定带来的困难，实现了“北斗三号”全球高精度导航定位授时服务。

上海天文台、紫金山天文台、南京大学等单位，深度参与了月球和火星探测等任务。在高精度 VLBI 测量基础上，发展了 S/X 波段 ΔDOR 测量技术和多探测器同波束 VLBI 观测技术，测量精度分别达到 0.2ns 和 ps 量级。综合利用测距测速和 VLBI 数据，月球和深空探测器定轨定位精度分别达到：环月段定轨 20 米，月面定位 100 米，巡视器相对定位米级，地火转移和环火段定轨分别优于 2 千米、100 米。

南京大学等单位构造了地月系统平动点附近的动力学替代轨道，系统研究了地月间转移轨道类型；分析并设计了航天器环绕、伴飞和着陆不规则小天体的轨道方案；系统研究了双小行星系统中的轨旋共振现象，提出了轨 – 旋耦合效应的快速计算方法。

改进了自主定轨算法，提出了卫星星座集中式自主定轨策略，可实现以星间链路为基础的自主定轨；改进了包括复杂摄动、适用各种轨道类型的空间碎片编目定轨算法；参与近地天体预警系统的建设，开发了近地天体轨道预报、初轨确定和精密定轨、撞击风险评估及主动防御的软件系统。

5. 天文历表相关工作

紫金山天文台在双星三星轨道拟合和恒星参数确定及经验质光关系、小行星高精度历表和质量测定、天然卫星历表和相关引力场参数测定、大行星数值历表表达和太阳系主要基频演化等方面都取得了重要成果。承担了我国天文历书服务，出版《中国天文年历》等多种天文历书，为海军研制天文导航软件，为国家海洋信息中心、天安门管委会等部门提供历书数据，为大众提供网络历书服务，制订了《农历编算和颁行》的国家标准。搭建了“太阳系天体高精度光学观测平台（国际编号 O49）”，所得自主资料在太阳系天体运动理论研究中发挥重要作用。

6. 相对论天体力学

上海工程技术大学和南昌大学构建了弯曲时空的显式辛算法；在后牛顿理论下构造了正则共轭旋转变量，揭示了同阶后牛顿拉格朗日与哈密顿一般不等价的关系，提出自洽后牛顿拉格朗日运动方程，为旋转体后牛顿动力学的可积判定和区分提供理论依据。紫金山天文台和南京大学建立了相对论等级式天文参考系体系，适合描述不同等级的天体系统和不同尺度的天文现象；研究太阳系二阶后牛顿光线传播及其测量，为高精度时频测量的科学应用提供理论基础。上海天文台给出了等效单体框架中相对论二体问题极端质量比的解析解，建立了极端质量比二体问题轨道演化数值方法，构建了时频传递相对论模型、相对论大地水准面并研究了光在缓慢运动天体附近的传播。

天文学史

孙小淳，王广超

天文学史作为科学技术史学科的重要方向，在过去十年中，取得了一系列重要的进展，具体表现在项目研究、学术交流和国际影响等三个方面。

中国天文学史研究保持了与天文学的密切联系，研究的问题从宇宙论到天象记录，从天文观测到天文仪器，从天文理论到历法计算，从星表到星图，从中国天文学传统到中外天文学交流，始终把握住了天文学史中的科学问题，因而得到天文学界的持续支持。近十年来，天文学史研究从国家自然科学基金获得了20多个面上项目的支持，使得天文学史研究能够持续积累和发展。

与此同时，天文学史在中国社会科学领域也获得了重要突破。近十年中，我国天文学史研究者从中国社会科学基金获得了3项重大项目。西北大学主持的《中国历法通史研究》对中国古代历法的算法进行“复原”研究，使得中国古代历法的算法逻辑更加鲜明，展示了中国古代历法的科学性质。中国科学技术大学主持的《中、日、韩古天文图的整理与综合研究》对散落在中国、日本、韩国、朝鲜以及欧洲和北美洲的中国古星图进行搜集、整理和数字化研究，为研究东亚各国之间的天文学交流做出了重要贡献。中国科学技术大学主持的《汉唐时期沿丝路传播的天文学研究》对汉唐时期传入中国的域外天文历法知识进行考证和梳理，研究域外来华天文学对本土天文历法的影响，呈现出一幅既宏观全面又不乏细节的汉唐时期中外天文学交流全景图。这些重大项目的立项，标志着天文学史研究在历史学领域也得到了广泛的认可，使之成为名副其实的科学与人文交叉性学科。

天文学史项目的组织促进了天文学史的研究，研究成果显著。近十年出版论著5种以上。其中：《诸史天象记录考证》（2015）应用现代天文计算方法，对“二十四史”及《清史稿》“本纪”和“天文志”中全部天象记录进行了考证和检验；《唐代域外天文学》（2019）考察了域外天文学与唐代的社会、宗教、政治、文化等因素之间的互动关系；《通天之学—耶稣会士和天文学在中国的传播》（2019），以耶稣会士和天文学为主题，在全球史和跨文化的视野下系统阐述天主教传教士与欧洲天文学传入中国的诸面相；《大统历法研究》（2020），在大量新史料的基础上，系统地介绍了与《大统历》有关的各种著作，揭示了《大统历》在明代的编修、使用及传播情况；《<崇祯历书>合校》（2017）是对明末中西天文学交流汇通研究的重要原始材料的整理和研究。此外，我国天文学史研究者在国内外学术期刊上发表重要学术论文40多篇。

我国的天文学史界特别注重学术交流。从2017年开始，中国科技史学会每年举行学术年会，天文学史专业委员会在每届年会上都举办天文学史专场，每次都有30多位学者参加，展示了一个十分活跃的学术共同体。此外，我国的天文学史研究在国际上也有很好的形象。2012年8月，第28届国际天文学联合会（IAU）大会在北京召开，天文学史专委会组织了《中国古代天文学成就展》，大会特邀科普报告《中国古代天文学》，不仅宣传了中国古代的天文学，而且进一步推动了天文学史研究。在此后的历届IAU大会上，都有中国学者做的邀请报告。2020年8月，在布拉格召开的第26届国际科技史大会上，我国天文学史研究者组织了多个专题研讨会。

近十年来，我国的天文学史研究在国际上的影响力也越来越大。2015年，在第29届IAU大会期间，我国学者当选为IAU天文学史专业委员会主席。2017年，我国学者当选为国际科技史学会（IUPHST / DHST）古代与中世纪天文学委员会主席。这些国际组织的任职，在国内都是首次。另外还有多名天文学史学者当选国际科学史学院（IAHS）院士或通讯院士，体现了中国天文学史研究的国际影响力。

天文教育

李向东

2012 年到 2021 年是中国天文教育发展最迅速的十年。截至 2021 年底，中国大陆共有 22 所大学开展天文学教育和研究，其中 12 所设立了天文学本科专业。表 1 总结了 2012 年以来新增的天文专业或天文系的情况。

表 1：2012 年以来中国大陆高校天文专业建设情况

序号	天文本科专业培养单位	专业设置年份	院系成立年份
1	厦门大学天文学系	2012	2012
2	河北师范大学空间科学与天文系	/	2012
3	云南大学天文学系	2014	2013
4	西华师范大学天文系	2015	2016
5	黔南民族师范学院物理与电子科学系	2015	/
6	中国科学院大学天文与空间科学学院	2016	2015

序号	天文本科专业培养单位	专业设置年份	院系成立年份
7	中山大学物理与天文学院	2017	2015
8	上海交通大学天文学系	2017	2017
9	贵州师范大学天文系	2017	2019
10	清华大学天文系	/	2019
11	华中科技大学天文学系	/	2019
12	广州大学天文学系	/	2020

此外，不少大学设立了天文学/天体物理学硕士点或博士点，如广西大学、湖南师范大学、华中师范大学、南昌大学、南京师范大学、山东大学（威海）、陕西师范大学、上海师范大学、天津师范大学、武汉大学、云南师范大学和中南大学等。

随着我国天文教育力量不断壮大，教育部于2013年成立了天文学类专业教学指导委员会，由于大部分委员与教育工作委员会委员重叠，近十年的天文教育工作是由两个委员会联合开展的。

一、制定《天文学类专业教学质量国家标准》

本科专业类教学质量国家标准是该专业类人才培养、专业建设等应达到的基本要求。主要用于三个方面：一是作为设置专业的参考；二是作为人才培养和专业建设的指导；三是作为质量评价的参考。

《天文学类专业教学质量国家标准》由南京大学于2013年牵头研制，历时4年多，期间组织了多次研讨和修改。2018年包含该标准在内的《普通高等学校本科专业类教学质量国家标准》由高等教育出版社出版。

二、制定《天文学类专业认证标准》

依据《普通高等学校基本办学条件指标（试行）》、《普通高等学校本科专业类教学质量国家标准》，南京大学于2018年牵头完成制定了《天文学类专业认证标准（第一级）》，即国家对普通高等学校天文学类专业办学的基本要求。

三、申报和建设天文学类一流本科专业

2019年教育部办公厅发布《关于实施一流本科专业建设“双万计划”的通知》，计划在2019–2021年建设10000个左右国家级一流本科专业点和10000个左右省级一流本科专业点。南京大学、中国科技大学、贵州师范大学、北京大学和北京师范大学入选国家级一流本科专业点。

四、申报和建设基础学科拔尖学生培养计划2.0基地

“基础学科拔尖学生培养试验计划”是教育部为回应“钱学森之问”而出台的一项人才培养计划，旨在培养中国自己的学术大师。2018年该计划实施2.0版，新增天文学等专业。南京大学、中国科学技术大学和北京大学分别于2019年、2020年和2021年入选基础学科拔尖学生培养计划2.0基地。

五、建设优质课程

2019年10月，教育部发布《教育部关于一流本科课程建设的实施意见》，计划经过三年左右时间，建成万门左右国家级一流本科课程和万门左右省级一流本科课程。南京大学五门专业课程入选国家级本科一流课程，一门课程入选教育部课程思政示范课程。

此外，多所高校在国内外慕课平台上开设天文学通识课和专业课，推动了全国高校天文教育的资源共享与优势互补，为广大学子提供了丰富的学习资源。

六、举办多场教育教学会议

举办或参与举办了多场教育教学研讨会议，开展全国天文公选课调研和新冠肺炎疫情期间线上教学情况调研，有力促进了高校之间的在人才培养计划、课程设置和教学方法方面的交流与合作。

天文普及

曹晨，朱进

由中国天文学会主办，北京天文馆、学会普及工作委员会等承办的“全国中学生天文知识竞赛”（2019年之前为全国中学生天文奥林匹克竞赛，由学会普及工作委员会和北京天文馆等单位发起和举办）每年举办一届，竞赛分预赛与决赛，决赛通过闭卷笔试、望远镜操作、实地观星三个环节决出金银铜牌及最佳成绩奖，在此期间还举办天文教师论坛活动，并进行国家集训队选拔赛，后确定代表中国参加各项国际天文奥林匹克赛事的选手。

2017年学会普及工作委员会创办“全国大学生天文创新作品竞赛（CAIC）”，竞赛由中国天文学会主办，学会普及工作委员会、教育工作委员会及决赛举办单位等承办，至今已成功举办四届。竞赛分作品征集、初赛评审、决赛答辩及作品展示、颁奖及研讨交流等环节，含天文科技与科普创新类作品，最终评出一、二、三等奖及各单项奖。上述竞赛体系为培养、提高青少年的天文兴趣与科学探索精神作出贡献，以期为我国天文科研、教育与科普事业培养储备优秀的后备人才。

由平塘县人民政府、学会普及工作委员会主办，北京天文馆、天津市天文学会等单位协办的“平塘县校园天文科普教学教师培训”于2017年8月在FAST所在的贵州省黔南州平塘县成功举行。培训旨在普及天文知识、提高公众天文科学素养、推进中小学天文教育和课程建设，培训内容包括教学所需的基础天文知识、经典天文实践活动、天文观测指导、天文资源包的教学使用等。当地45名天文教师参加了培训，在当年推动当地7所学校约4000名中小学生开始学习天文校本课程，截至2020年，当地已有近60所中小学校引入天文校本课。此外，学会普及工作委员会还多次举办全国科技教师天文知识与技能培训暨天文科普教育论坛活动（2016年9月于延庆、2017年10月于贵州等），于2016年5月在新疆乌鲁木齐市举办了天文业务培训班暨天文研讨会。

于2019年全年开展了庆祝国际天文学联合会成立100周年的“IAU100”系列天文科普活动，活动由中国天文学会主办，学会普及工作委员会、北京天文馆等承办，活动主题为“同一天空下”。开展的主要活动包括：2019 年中国大陆地区 IAU100 启动仪式，于 2019 年 1 月 12 日上午在北京天文馆举行；IAU100 太阳系外行星世界命名（NameExoWorlds）——中国（内地）活动，于 2019 年 5–11 月开展，并于 12 月 21 日上午在北京天文馆召开命名结果媒体说明会，经专家评审、公众投票及 IAU 核准确认，最终由我国天文学家发现的首颗太阳系外行星 HD173416b 及其母恒星正式命名为望舒与羲和（提名来自广州第六中学天文社）。

每年均召开学会普及工作委员会工作会议，参与组织中国天文学会学术年会“天文学史、教育与科普”分会场报告交流；于 2019 年 5 月 29–31 日在河北师范大学召开了“礼赞共和国、追梦新时代——天文科普志愿服务行动”座谈会，举行中国天文学会天文科普志愿服务行动启动仪式，举办了“情系星辰大海——中国天文科技工作者风采”图片展。

学会普及工作委员会组织号召委员所在单位、学会团体会员单位等积极开展各类天文科普活动，积极参加全国科技活动周、全国科普日等活动，在疫情与“双减”政策等影响下积极探索天文科普的新方向与新思路，开展线上科普工作。于 2020 年举办了“仰望星空”科学之夜和科技活动周重大示范活动——“一带一路”科普交流周，于 2021 年 5 月参与举办了“中国天文学会空间天文探测科普”学术交流活动，开展了 2021–2025 年度全国科普教育基地申报推送工作。

天文学名词审定

余恒

2013年8月，《海峡两岸天文学名词》正式由科学出版社出版，并在同年10月天文学名词审定委员会（以下简称“天文名词委”）第六届第三次会议上举行了简朴的首发式。这项工作始于2004年，经过两岸天文界同仁近10年的不懈努力，终于问世。这是海峡两岸天文学家加强交流的结果，也是“海峡两岸天文学名词工作委员会” 取得的初步成果。它为两岸天文学在学术交流、科研合作，以及科学教育、普及和传播上提供了方便，也为今后两岸天文学名词进一步一致化奠定了扎实的基础。

随着名词数据库的不断完善和新词的持续增补，网站内容与原有纸本天文词典之间的差别越来越大，再版天文名词词典的需求愈发迫切。在多方努力下，新版《英汉天文学名词》终于获得中国科协三峡科技出版资助计划支持，于2015年12月由中国科学技术出版社出版。新版天文学名词词典采用程序自动排版，与数据库保持完全一致，在很大程度上避免了人工录入排版造成的差错。新版词典收录含词条约26700个，与2000年出版的天文词典相比，增补词条约6000个、注释3000多条；修正排版错误、录入错误，规范大小写、连字符用法，移除不规范名称等共2100余处。新版词典在同年的天文学年会上作为会议资料发放，极大地方便了国内天文专业学生和科研工作者开展相关工作，很好地宣传了天文名词委的审定成果，扩大了名词工作的影响。

在新版天文学词典出版后，天文名词委的工作重点集中到第三版《天文学名词（定义版）》上。作为第一个出版学科名词定义词典的委员会，天文名词委计划在原有1998年《天文学名词（定义版）》基础上，将条目总数翻倍，达到4500条，以充分体现天文学科的前沿成果和最新进展。定义版的编撰由10个子学科分头进行。经过几年深入细致的工作，新版《天文学名词（定义版）》总条目实际达到4873条，超过预定目标近400条。试用稿已于2021年9月向社会公布，预计于2022年正式出版。

除词典编撰工作之外，天文名词委也一直在努力提升公共服务质量和用户体验。2017年，重新设计制作的新版英汉天文学词典取代了运行多年的旧版系统正式上线，并纳入中国虚拟天文台平台。新版网站在安全性、可靠性、适应性、运行效率、用户友好度等许多方面都有大幅提升。

此外，为助力我国的火星探测任务，服务社会公众，天文名词委在2020年组织力量首次将国际天文学联合会（IAU）已公布的全部1950条（截至2020年7月15日）火星地形地貌名称译为中文，经相关领域专家审定后，将全部数据向社会开放使用。这为相关学科的研究和科普教育等活动提供了有益的参考。而且国家天文科学数据中心为此次发布专门制作了三维可视化页面，能够直观地显示火星地名在火星地图上的实际位置，为火星地名的辨识与研究提供了便利。

天文期刊

侯金良等

《天文和天体物理学研究》(*Research in Astronomy and Astrophysics*，缩写 *RAA*)是2009年由《中国天文和天体物理学报》(缩写 *CHJAA*)更名而来，是我国天文领域唯一的SCI国际天文期刊(月刊)，国际化程度高，具有一定的国际影响力。

在过去的十年，*RAA* 积极服务于我国天文重大科学计划，先后发表了一批基于我国天文大科学工程的重要成果，包括探月专刊，LAMOST科学成果专刊，FAST大型仪器设备成果专刊，先基太阳天文台和中国天文选址专刊等。同时 *RAA* 也积极开拓国际合作，发表了一批具有国际重要影响力的成果，包括IAU大会专刊、美国30米望远镜(TMT)科学白皮书、纪念哈勃望远镜30周年及系外行星等综述性文章。在2012年IAU会议期间，*RAA* 承办了IAU期间的官方报纸，并组织承办了一系列活动。

十年来，*RAA* 连续两次获得中国科技期刊国际影响力提升计划B类资助，获得国家新闻出版广电总局百强报刊奖，获得第四届中国出版政府期刊提名奖等。在国家政策的鼓励下，近年来 *RAA* 收稿和发稿量已大幅提高，相比2012年，*RAA* 投稿以及发文数目增长近2倍，但整体稿件质量和国际一流期刊的距离依然不小。

《天文学报》已走过60多年的发展历程，涵盖天文学各领域。经过第十届、十一届、十二届编委会和编辑部共同努力，在2012–2022年期间取得了长足的进步。期刊影响力保持上升趋势，平均出版周期不断缩短；先后增加了博士学位论文摘要选登、特邀评述、天文大科学工程、天文教育、热点评论专栏；投审稿平台升级，开通期刊微信公众号，探索数字出版新模式；先后获得“百种中国杰出学术期刊”、“中国精品科技期刊”、“期刊数字期刊影响力100强”、“江苏省新闻出版广电政府奖(报刊奖)” 等荣誉，刊发论文入选 “第二届中国科协优秀科技论文遴选计划”。

《天文学进展》创刊于1983年，乘改革开放之风，一路走来，见证了中国天文学近40年的发展历程。过去10年来，为了适应新形势下的天文学发展，进展论文范围从过去的只有天文学前沿领域综述文章，发展到现在的综述和研究论文并重。同时，刊登论文的范围不再局限于基础天文观测研究和天文技术领域，深空探测和导航定位等应用方面的文章同样受到重视。此外，编辑部校编质量得到提升，在上海市和中科院有关部门组织的检查中，总体质量均获得优秀。

《天文研究与技术》由《云南天文台台刊》（1977 年创刊）更名而来，2004 年正式出版。主要刊登天文学及相关学科的观测研究、实测技术和方法、专题述评等方面的学术论文，以及有关天文新发现的研究快讯。2015 年，《天文研究与技术》网站全新上线，启用新的投审稿系统，并上传了从 1977 年创刊至今的全部论文。先后被包括中国科学引文数据库（CSCD）以及国际天文数据库 ADS 和 CDS 等在内的数据库系统收录。2022 年《天文研究与技术》将由季刊变更为双月刊。

天文图书

侯金良等

中国科学院国家天文台图书馆位于国家天文台总部，前身是北京天文台图书馆。国家天文台图书馆围绕天文台科研工作规划文献情报服务。支撑天文台前沿天文基础科学研究，支撑500米口径球面射电望远镜（FAST）、大天区面积多目标光纤光谱望远镜（LAMOST）、“天问一号”、空间站工程巡天空间望远镜（CSST）等多个国家大科学装置研发运行使用。目前馆藏6957册天文物理相关藏书，2795册（合订本）天文物理相关期刊。

中国科学院紫金山天文台图书馆最早可以追溯到1935年，是我国馆藏资源最为丰富的天文学图书馆之一，也是东亚地区最大的天文学专业图书馆之一。馆内藏有图书和期刊（合订本和单行本）300000余册，包含了天文领域多种从创刊开始的出版物，是中国科学院的特色馆藏之一。紫金山天文台图书馆秉承深厚的文化底蕴，坚持人性化的服务理念，以数字图书馆建设为方向，全面推进图书馆的现代化建设，调整资源结构配置，丰富馆藏资源，强化资源共享，在科学研究、人才培养和社会服务中发挥应有作用，努力实现主动、精准、泛在的知识服务，向现代化图书馆不断迈进。

中国科学院上海天文台图书馆源于19世纪末法国天主教会创办的佘山天文台和徐家汇天文台的图书馆，馆藏有一百多年的天文专业书籍和期刊，是国内收藏天文期刊最为古老的图书馆。目前馆藏中外文书刊60000余册，自建特色数字文献库馆藏天文类图书近15000本，每年提供各类型情报产品90余份，开展各类型讲座活动50余场。

中国科学院云南天文台图书馆成立于1972年。从成立以来，图书馆一直致力于为科研服务。随着网络应用的普及，图书馆把工作重点转移到电子资源的保障。目前馆藏资源包括英文书7921册，会议文集561册，工具书1205册，天文年历343册，中文图书9737册，中文期刊2011册。

中国科学院国家授时中心图书馆，最早名为中国科学院陕西天文台图书情报研究室，成立于1966年。国家授时中心图书馆目前印本馆藏量为7.6万余册，主要涉及时频、导航、物理、天文、通信和计算机等学科。其中馆藏较全的会议录有全国时间频率年会论文集、中国卫星导航学术年会论文集、ION、PTTI、EFTF、IFCS 等会议文集。拥有 IEL、APS、Springer 自购数据库以及全院开通的其他可利用的数据库资源共 118 种。

新疆天文台图书资料室面积约 120 平方米，拥有天文专业图书 8000 余册、期刊 38 种 10000 余册。它既是本台科研人员必需的一个服务平台，又是国家天文台科技文献服务的有力支撑。2011 年开始，该台图书文献室加入中科院统一自动化系统，逐步形成了新疆天文台特色的知识化服务能力，建立起新时期研究所文献情报服务模式。

随着信息技术的高速发展，适应科学家文献资源检索、获取的方式变化，完善资源保障，科学院各天文单位图书馆大力发展数字图书馆的建设，数字图书馆囊括了 Springer-Nature、Science、IOP、Elsevier、APS、IOP、SPIE 等多个出版社上万种天文、物理、光学等资源，并围绕科研过程形成了针对不同研究领域的知识服务中心和情报分析服务体系。

天文信息

崔辰州，李珊珊

相对其他委员会，中国天文学会信息化工作委员会（简称“信息委”）无疑是年轻的。2019年3月25日，经过中国天文学会十四届理事会第一次会议、常务理事会第一次会议批准和委员会内部严格筛选，信息委首届共计42名委员名单完整出炉。信息委的职责主要包括：1. 定期组织学术交流活动；2. 组织制定天文信息化相关技术标准和管理标准；3. 组织评选“信息化工作先进个人”；4. 代表中国天文学会对接国内外信息化有关部门、机构、团体组织；5. 承接中国天文学会交办的其他事宜等。2019年6月14日，信息委成立大会暨信息化工作交流研讨会在南京召开。中国天文学会信息委宣布成立，同时为信息委官方网站揭幕。委员们集体审议了信息委章程，并以自愿报名的形式分别加入科研信息化工作组、管理信息化工作组、科教信息化工作组、信息化基础设施工作组，将结合自己的实际工作为中国天文信息学的全方位发展出谋划策，贡献自己的力量。此后，信息委第一届委员会第二次工作会议、第三次会议于2020年和2021年分别在厦门、昆山召开。会议上除专题报告、工作总结外，分别结合年度工作及当前天文学信息化工作的热点、难点话题展开讨论，并就充分发挥委员会集体力量组织好虚拟天文台与天文信息学年会、积极组织制定信息化相关标准规范、信息委年度重点工作事项、校园虚拟天文台联盟建立等议题展开充分讨论。信息委的成立，是中国天文信息学发展的里程碑。

2019年6月5日，科学技术部办公厅印发了《科技部、财政部关于发布国家科技资源共享服务平台优化调整名单的通知》，以国家天文台为依托单位，紫金山天文台、上海天文台、中科院计算机网络信息中心为共建单位申报的“国家天文科学数据中心”被正式列入国家科技资源共享服务平台名单。同年10月，中科院国家天文台、丽江市政府、中科院云南天文台、云南大学四方签署合作框架协议，在丽江建设国家天文科学数据中心丽江分中心。2021年3月，天津大学、广州大学与国家天文科学数据中心签订合作共建协议，将分别承担国家天文科学数据中心技术研发创新中心和国家天文科学数据中心粤港澳大湾区分中心的建设任务。2021年12月，国家天文科学数据中心与华中师范大学、河北师范大学、西华师范大学签订合作协议，共建教育研发应用中心，协同实施基于真实科学数据的课程研发、教育教学实践，实现科学数据、信息技术与教育教学内容深度融合。

信息委成立至今，所有委员积极推动天文信息学这个同样年轻的学科在中国的发展。自2019年3月起，多次举办研讨会及技术培训、讲座，累计超万人，在北京、上海、辽宁、新疆、大庆等地多次举办线上线下的以天文数据为核心驱动的多种多样的科普教育活动，如：2021年上海天文馆正式对外开放；万维望远镜宇宙漫游大赛已举办五届，历经十余年；新疆天文台创办“新疆青年科普行”系列活动已举办40多期。这些活动及项目都具有一定的社会影响力，推动了天文学的科普教育发展。

信息委的宗旨是团结、联合、组织天文信息技术相关领域的专业人士，开展学术技术交流和战略研究，制定相关标准，组织专业培训等，提高我国天文学领域科研信息化和管理信息化研发与应用水平，提升国际影响力。未来，信息委将继续围绕这一宗旨并结合当前的实际形势展开工作，不忘初心，行而不辍，争取继续从天文科研、管理、科教、基础设施等方面推进中国天文信息学及天文学的发展。

青年天文工作者

田斌，安涛

鉴于国内天文科技工作者队伍呈现出明显的年轻化特征，为了帮助青年科技工作者构建良好的发展环境，使他们安心治学，促进他们成才，2016年1月，中国科学院各天文系统单位和国内部分高校青年天文工作者共同发起，向中国天文学会提出了设立青年工作委员会的建议。经中国天文学会理事会研究，批准设立中国天文学会青年工作委员会，并作为中国天文学会下设的工作委员会之一。青年工作委员会负责统筹协调青年天文工作者培养，促进青年天文工作者学术交流和学科交叉等工作。

为了规范和推进相关工作，委员会于2016年11月在北京举行了工作会议。重点讨论了青年工作委员会实施细则，中国天文学会“青年天文论坛”组织情况，国内青年天文学者的学术交流活动策划方案，以及青年天文学者参与科学普及工作的原则和思路等内容。为了进一步扩大青年工作委员会的影响力，使其具有更广泛的代表性，委员会于2018年1月在云南昆明举行第二次工作会议，重点讨论了增补若干高校青年天文学者的委员的方案等内容。

自2012年开始，以中国科学院各天文系统单位青年创新促进会小组为核心，国内青年天文工作者联合创办了“青年天文论坛”，并于2014年正式列入中国天文学会系列学术会议之一。青年工作委员会成立后，论坛由委员会负责主办，逐步打造成为了国内青年天文学者合作交流的品牌平台。

为了进一步发挥国内青年天文学者在天文学科发展规划方面的积极作用，青年工作委员会联合中国科学院相关部门于2018年1月19–21日在云南昆明举办了“青年天文学者研讨会”，会议由中国科学院云南天文台青年创新促进会小组和中国科学院天体结构与演化重点实验室承办。这次会议旨在倾听国内青年天文学者对“十四五”（2020–2025）及以后中国天文发展的想法和建议，开始为“十四五”及以后的发展规划做准备，讨论中国天文应该重点发展的前沿领域、资源配置的优化、设施装置的建设以及实施方法等。会议筹备期间通过15个分学科的微信群共收到建议1万余条，通过电子邮件收到建议60条，发展报告

34个。33位会议召集人负责了建议的搜集和整理工作，并形成会议报告共48个，其中综述类报告12个，学科发展建议报告36个。来自国内28家单位的130余位天文学者参加了会议。会议还开通了网络视频直播平台，有600余人次通过视频直播平台观看了会议直播。会议对于中国未来大型科学装置的建造和运行开展了深入、热烈的讨论，为青年一线科学家向相关大型科学项目献计献策提供了极为宝贵的渠道。

青年工作委员会成立以来，在促进青年天文学者成长和交流等方面开展了一系列工作，青年学者通过委员会搭建的平台，增进了解，互相学习，共同进步。青年学者是科学研究的生力军，是天文学科发展的未来。青年工作委员会在后续工作中，将继续打造好已有的交流平台，并深入到青年学者中去，积极收集、了解青年天文工作者面临的发展问题和建议，制定相应的工作方案，为青年学者成才发挥更加积极作用。

女天文工作者

王娜

2017年8月，女天文工作者委员会正式成立。作为中国天文学会下设的工作委员会之一，女天文工作者委员会的主要职责为：一、促进女天文工作者的成长与发展，为女天文工作者的学术提升提供帮助，推动女天文工作者全面参与天文事业的发展；二、发现与宣传优秀女天文工作者，展示女性科技人才的风采，提升女天文工作者在科技创新和社会发展中的作用和地位；三、组织学术交流活动，协助女天文工作者扩大科研合作，为女天文工作者展示才华与成果搭建平台；四、收集、了解女天文工作者面临的发展问题，反映女天文工作者的建议、意见和诉求，并向有关部门、单位反映并提出建议。中科院新疆天文台王娜研究员担任首届女天文工作者委员会主任。

2018年10月27日，召开了首届中国天文学会女天文工作者交流会议，共有70余位专家学者参加。中科院紫金山天文台杨戟研究员致辞并宣读了《中国天文学会女天文工作者委员会工作实施细则》；崔向群院士介绍了中国天文学界历届优秀女性天文学家；云南天文台王晓彬研究员分享了她多年从事小行星研究的经验和体会；国家天文台梁艳春研究员介绍了IAU女天文工作者工作组的情况。参会人员还就女天文工作者在科研、家庭和工作等方面面临的问题进行了深入交流和讨论，大家一致认为："在追逐星空梦的道路上可能会因为家庭琐事消退你的激情，磨灭你的理想，这时希望广大女天文工作者，特别是拥有广阔发展空间，承载着伟大时代使命的年轻天文工作者能够以老一辈的筑梦人为榜样，秉承着对天文的热爱和执着，怀揣着对实现梦想的期待和希望，不忘初心，坚定信心，传承匠心，敢于啃难啃的'硬骨头'，以女性特有的坚韧和毅力，以严谨务实的科研态度潜心研究，探索创造出属于自己的一片天。相信在不久的将来，会有更多优秀的女天文工作者活跃在一线的科研工作中。"

2019年12月26–27日，第二届女天文工作者交流会在新疆天文台南山观测站顺利召开。来自国内天文机构和高校近70位专家、学者和学生参加了该会议。中科院国家天文台汪景琇院士，通过列举国际上女天文学家贡献案例，对比国内外天文学文献数据，客观分析了当前中国天文学面临的机遇与挑战。中科院国家天文台南京天光所崔向群院士以国际上天文望远镜发展历程和前沿技术为背景，介绍了我国光学望远镜发展现状，鼓励大家把握最好发展时期。风采绽放谱芳华。与会代表结合各自工作方向进展作报告，展现了活跃在天体物理、天文技术、空间目标与碎片等多个前沿领域的靓丽身影。广大女天文研究工作者孜孜不倦、卓越追求，以新时代中国科学家责任标准要求自我，用自己的努力撑起美丽星空，为中国天文事业的发展发挥了不可替代的作用。

由于疫情，每年一次的女天文工作者交流会暂停两年。虽然疫情限制了女天文工作者交流会的召开，但每一位女天文工作者一直在各自的工作岗位上发着光和热，默默地为中国天文学事业的发展做出应有的贡献。

天文学会

宁宗军

2022年，中国天文学会迎来百年华诞，和中国共产党年龄相当。按照每十年为一个阶段的划分，须回眸刚过去十年（2012–2021年）中国天文学会的历程。在中国科协的领导下，中国天文学会依靠各团体会员单位、各省市学会以及广大会员的大力支持，面向世界科技前沿，面向国家重大需求，在建设国家重大工程、组织和促进学术交流、加强国际合作、开展科普宣传、加强组织建设等方面，不断探索创新发展，成绩斐然、硕果累累。如今，中国天文学会已经成为代表全国天文科技工作者的最重要的社会团体，党和政府联系科技工作者的桥梁和纽带。

一、搭建国内天文界一流学术交流平台

十年来，中国天文学会把立足天文相关学科发展前沿、举行形式多样的学术交流活动作为学会最主要的任务。学会力争每年主办一次全国性学术年会，力争打造成为国内天文一流学术交流会，同时协同各专业委员会、工作委员会共同举办多场次国内外学术会议。据统计，十年来共有学术活动70次，参会代表超过100000人次，会议报告论文有4000多篇。

十年来，中国天文学会年度学术年会分别在苏州、西安、乌鲁木齐、昆明、德令哈、北京、南充等地举行七次，参加学术年会的代表来自国内100多个单位，共计近9000人次，交流报告2600多篇，大会特邀报告100多篇，面向公众的高级科普报告和高级天文讲座40篇。近两年受新冠疫情影响，2020和2021年学术年会采用线上和线下结合方式举行。

二、加强国际合作与交流，增强我国天文学的国际地位，积极响应IAU100活动

中国天文学会及其会员单位积极开展同国际天文联合会(IAU)学术交流活动，积极推荐IAU会员和参与IAU主导的国际活动。近十年来，中国天文学会组团参加了第29届（2015年在美国夏威夷州）、第30届（2018年在奥地利维也纳）活动，共推荐IAU会员80余人，他们是来自国内科研院所、高校等第一线的天文工作者。下一

届（31）届 IAU 大会计划于 2022 年 8 月在韩国召开。第 30 届大会的亮点之一是 IAU100 项目的启动。2019 年是国际天文学联合会成立 100 周年（IAU100）。为纪念这个划时代的事件，IAU 将组织贯穿一年的庆祝活动，通过宣传一个世纪以来的天文发现，增加公众对天文学作为一个工具在教育、发展以及对外交流方面的重要影响的认知。IAU100 的主题是“同一天空下”。2019 年中国大陆地区 IAU100 启动仪式由中国天文学会主办，中国天文学会普及工作委员会和北京天文馆承办，于 2019 年 1 月 12 日上午在北京天文馆举行。2019 年举行的 IAU100 全球庆典项目包括：在 1 月 10 日至 13 日举行“天文学 100 小时”全球观星活动；在爱因斯坦学校的旗帜下，建立一个学习引力和广义相对论等主题的全球校园网络；2 月 11 日 IAU 妇女和女孩天文日，在联合国国际妇女和女童科学日当天举办天文活动，鼓励更多女性特别是女孩积极参与；利用现有的教育项目并组成新的 IAU 暗夜大使网络，使暗夜保护的观念深入人心；巡回展览，通过过去一个世纪以来科技和文化的进步，来介绍现代天文学和空间探索的重要成就；2019 年 7 月登月 50 周年纪念活动；提高教师在科学课题和教学技巧方面的能力的行动；新版的 ExoWorlds 竞赛将为世界上所有国家提供命名系外行星的机会；在 2019 年 7 月 2 日日全食期间，庆祝日食观测验证爱因斯坦相对论百年，等等。

三、积极开展科普活动和中学生天文竞赛

天文学会的宗旨中就有“通俗天文学之普及”的目标。中国天文学会依托普及工作委员会与各省市天文学会、协会或天文小组、团体会员单位积极推动天文科普活动。不完全统计，2012-2021 年间共计科普讲座 200 多场，参加人数将近 30 万人次，科普宣传 400 次，青少年科技竞争赛 30 余次，组织青少年冬、夏令营 150 次，参加人数达上万人。典型的科普活动有 2014 年天文科普首次进佤山主题活动、2016 年第八届天文科普教育论坛、2017 年天文业务和教师培训班、2017-2021 四届全国大学生天文创新作品竞赛、2019 年全国中学生天文知识竞赛决赛、2020 年“全国科技工作者日”活动、2021 年主题为“微观世界探秘之旅”的第十七届公众科学日等系列活动。创办于 2017 年的全国大学生天文创新作品竞赛（简称为 CAIC 竞赛）是我国唯一全国性大学生天文类赛事。另外，2021 年 7 月上海天文馆落成并对外开放，这是全球最大的天文馆。

中国天文学会在新疆乌鲁木齐、云南昆明、北京、山东威海、山西太原、新疆伊宁、浙江绍兴、贵州、河北廊坊等地组织全国中学生天文奥林匹克竞赛十次，组织中国队赴巴西、韩国、孟加拉国、立陶宛、印度尼西亚、罗马尼亚、俄罗斯、鞑靼斯坦共和国、保加利亚、印度、泰国、斯里兰卡、

中国（北京、威海、丽江）、匈牙利等参加国际天文竞赛，获得39枚金牌、67枚银牌、26枚铜牌。

四、中国天文学会机构建设

十年来，中国天文学会经过第十二届、十三届和十四届理事会，共换届两次，分别在2014年和2018年召开全国会员代表大会。共召开常务理事会或者理事会共50多次，共同研究和制定学会工作及方案。贯彻落实党的十九大精神，加强学会党组织的建设，2016年成立中共中国天文学会党委，2018年党委换届，共召开党委及扩大会议10多次，集体学习和研讨30多份文件。2016年10月31日成立青年工作委员会，2017年8月7日成立女天文工作者委员会，2019年3月25日成立信息化工作委员会。至此，中国天文学会共有8个工作委员会和基金工作组，以及原有的11个专业委员会。

为纪念陆埮先生，促进中国天文学的发展，鼓励并表彰在天文和天体物理研究中取得重要原创性成果的青年学者，中国天文学会十四届六次常务理事会决定中国天文学会特设立“陆埮奖”，每两年授予一次，每次获奖人不超过两位，2022年是第一届陆埮奖评选，奖金由南京紫金山天文台小行星基金会支持。

天文大事记
（2012—2022）

2012 年

65 米射电望远镜（天马望远镜）落成，成为亚洲最大的单口径全可动射电望远镜。

2012 年 1 月 16-20 日

国际电信联盟（ITU）2012 年无线电通信全会（RA）在瑞士日内瓦举行。我国建议书的立场是不赞成目前取消闰秒，维持现行的国际标准时间—协调世界时（UTC）的定义不变，该建议赢得了大多数国家的支持。

2012 年 1 月 20 日

贵州省机构编制委员会办公室正式批复成立贵州射电天文台。

2012 年 2 月 8 日

中国第 28 次南极考察队结束科考。南京天文光学技术研究所成功安装了我国首台可远程控制自动跟踪的天文光学望远镜 AST3-1。

2012 年 2 月 22 日

北京市人民政府天安门地区管理委员会向紫金山天文台赠送国旗，紫金山天文台天文历算研究人员按时为天安门管委会提供下一年度北京天安门地区每天日出日落的标准时刻。

2012 年 4 月 11 日

由南京天文光学技术研究所为南京大学研制的光学近红外太阳色球爆发探测仪（ONSET）试观测并获得国内首张红外波段太阳全日面像和局部像。

2012 年 5 月 20 日

海峡两岸陨石专家鉴定福建省福清东翁 585

公斤陨铁，此为我国私人藏家手上质量最大的陨铁标本。

2012年5月21日

我国东南沿海出现日环食，国内外多支天文爱好者队伍莅闽。

2012年6月1日

厦门大学正式决定复办天文学系。

2012年7月4-6日

第429次香山科学会议——“天体物理新视野与大口径射电望远镜”研讨会在乌鲁木齐召开。

2012年7月15日

新疆奇台天文观测及科普教育基地奠基仪式举行。

2012年8月25-31日

IAU第28届大会在北京成功举办，时任国家副主席习近平出席开幕式并致辞。这是我国参加IAU70多年以来第一次举办国际天文学联合会大会。大会期间IAU天文推动发展东亚区域办公室在北京成立。

2012年8月31日

北京大学刘晓为教授当选国际天文学联合会副主席。

2012年10月

紫金山天文台青岛观象台被评为2012-2016年“全国科普教育基地”。

2012年10月8日

紫金山天文台主导的“暗物质粒子探测卫星”有效载荷顺利完成在欧洲核子中心（CERN）的束流试验。

2012年10月15-16日

“十月天文论坛－中国天文的过去、现在和未来－暨庆贺王绶琯先生九十华诞典礼”在北京举办。

2012年10月26-27日

“天文科普入寺院、医疗健康进藏区”活动在香格里拉成功举办。

2012年10月28日

上海65米射电望远镜落成仪式及上海天文台成立50周年暨建站140周年庆典活动在松江佘山上海65米射电望远镜现场隆重举行。

2012年11月19日

紫金山天文台与荷兰代尔夫特理工大学（TU Delft）科学与技术合作谅解备忘录签约仪式在南京举行。

2012 年 11 月 26 日

举办厦门大学天文学系复办仪式。

2012 年 12 月

天马望远镜第一次加入“嫦娥二号”任务观测。

2012 年 12 月 13 日

“嫦娥二号”卫星成功飞越小行星“战神”图塔蒂斯，在国际上第一次拍下这颗小行星的光学图像。这是我国首次实现对小行星的飞越探测，也是国际上首次实现对“战神”的近距离探测。

2013 年

天马望远镜参与完成“嫦娥三号”VLBI 实时测定轨，和后续历次探月任务和首次探火任务“天问一号”的 VLBI 测定轨。

2013 年

国家天文台汪景琇研究员当选为中国科学院数理学部院士。

2013 年 1 月 10-13 日

第十三届东亚地区亚毫米波接收机技术研讨会在南京召开。

2013 年 1 月 22 日

国家天文台新一代厘米 – 分米波射电日像仪成功获得第一张 1.7 京赫兹太阳射电图像。

2013 年 3 月 27-29 日

BigBOSS（DESI 前身）合作会议召开，上海交通大学天文系于 2018 年 4 月正式成为 DESI（Dark Energy Spectroscopic Instrument）项目的系级 / 所级成员单位。

2013 年 4 月

中国科学院行星科学重点实验室成立于上海天文台和紫金山天文台。

2013 年 4 月 27 日

中国科学院月球与深空探测重点实验室成立，挂靠国家天文台。

2013 年 5 月 26-29 日

2013 南极巡天望远镜（Antarctic Survey Telescopes, AST3）国际合作会议在云南腾冲顺利召开。

2013 年 6 月

国家天文台射电天文实验室（JLRAT）加入 SKA 天线（DISH）和宽带馈源（WBSPF）工作包联盟。

2013 年 6 月 20 日

国际天文联合会主席、日本国立天文台海部宣男教授应邀访问紫金山天文台。

2013 年 6 月 20 日

国际天文学联合会主席海部宣男教授访问南京天文光学技术研究所。

2013年8月

福建省天文学会与高雄市天文学会合作办理“闽台青少年天文夏令营”。

2013年9月3-6日

第五届海峡两岸天文望远镜及仪器学术研讨会在云南西双版纳成功召开。

2013年9月9-13日

第7次国际太阳偏振研讨会在昆明举行。

2013年9月10日

在“大型射电望远镜科学与技术国际研讨会”期间，新疆天文台与美国国家射电天文台在乌鲁木齐签署了《关于新疆奇台110m射电望远镜项目在射电天文领域内科学技术合作的谅解备忘录》。

2013年9月25日

中国天文学会开展“第四届黄授书奖”评选活动，上海交通大学杨小虎获“第四届黄授书奖”。

2013年10月

南京天文光学技术研究所崔向群院士荣获2013年度“何梁何利基金科学与技术进步奖”。

2013年10月7日

“中科院南美天文研究中心”在智利大学成立。

2013年10月7-12日

国际会议“Energetic Particles in the Heliosphere（日球中的高能粒子）”在南京举行。

2013年10月23日

《海峡两岸天文学名词》由科学出版社正式出版发行。

2013年10月27-31日

中国天文学会2013年学术年会在江苏苏州召开，参会代表500多人，大会特邀报告10篇，2个高级科普报告，8个分会场，交流口头报告298篇。

2013年11月10日

第十届中澳科技研讨会“天文和天体物理”在南京开幕。会议由中国科学院和澳大利亚科学院、澳大利亚技术科学与工程院联合主办。

2013年12月

国家授时中心“时间网站”被评为“2013年度中国科学院优秀科普网站”。

2013年12月10-12日

南京天文光学技术研究所成功举办首届“南极星”青年科学家论坛。

2013年12月14日

落月成功，我国实现第一次月面软着陆。

VLBI 分系统在落月后开展了对“玉兔”月球车的相对定位。

2013 年 12 月 28 日

中山大学天文与空间科学研究院在珠海校区举行了揭牌仪式，复办了中山大学的天文学科。

2013 年 12 月 31 日

新一代厘米 - 分米波射电日像仪建设竣工验收。

2014 年

厦门大学天文学系与国家天文台签署共建协议；与上海天文台签署共建协议，成立“上海天文台 - 厦门大学天体物理联合中心”。

2014 年 1 月

由南京天文光学技术研究所自主研制的南极视宁度监测仪 DIMM 在南极泰山站正式开始观测。

2014 年 1 月 22 日

大口径全天巡视望远镜中国团队（Large Synoptic Survey Telescope–China Consortium） 与 LSST Corporation 签署了合作备忘录。

2014 年 3 月 1 日

国家科学基金重大项目“LAMOST 银河系研究”启动。

2014 年 5 月

南京天文光学技术研究所为泰国 2.4 米望远镜研制的阶梯光栅光纤光谱仪顺利通过检测验收，该光谱仪于当年 10 月顺利完成现场装调并首光。

2014 年 5 月

青岛观象台被山东省科协、山东省财政厅命名为：四星级山东省科普教育基地。

2014 年 5 月 6 日

TMT 国际天文台有限责任公司（TIO LLC）成立，成员包括美国加州大学、加州理工学院、日本国立天文台、国家天文台、印度科技部、加拿大大学天文研究联盟。

2014 年 6 月 23-27 日

第 22 届国际天文馆学会（IPS）大会在北京举行。

2014 年 7 月

南京天文光学技术研究所研制的“超级”自适应光学系统（Extreme Adaptive Optics，简称：Ex–AO）作为客座仪器，成功对接欧洲南方天文台（ESO）3.6 米新技术望远镜（NTT），圆满完成了测试观测。

2014 年 7 月 17-18 日

JLRAT 实验室成员协助科技部在北京香山饭

店召开了以“中国射电天文学发展与平方公里阵列望远镜(SKA)”为主题的第S21次香山科学会议。

2014年7月22日

中国－阿根廷合作40米射电望远镜项目获得阿根廷科技创新部（MinCyT）正式立项。

2014年8月

新疆天文台申报的“110米大口径全可动射电望远镜（QTT）关键技术研究”项目（国家重点基础研究发展计划）成功立项。

2014年8月21日

中国科学院空间天文与技术重点实验室和中国科学院计算天体物理重点实验室成立，挂靠国家天文台。

2014年9月11日

中国天文学会开展“第十二届（2013-2014年度）张钰哲奖”评选活动，北京师范大学何香涛获“第十二届（2013-2014年度）张钰哲奖”。

2014年9月23日

“国家天文台·贵州师范大学天文研究与教育中心暨FAST早期科学数据中心”揭牌（在贵州师范大学）。

2014年10月

中国科学院南京天文光学技术研究所王亚男研究员荣获2014年度“何梁何利基金科学与技术进步奖”。

2014年10月15日

南京天文光学技术研究所与天仪公司共同完成印度太阳色球望远镜的现场安装与调试。

2014年10月27日

中国天文学会2014年学术年会暨第十三次全国会员代表大会在陕西省西安市临潼区隆重开幕。会议代表131人，选举产生了第十三届理事会，国家天文台武向平任理事长。

2014年10月29日

紫金山天文台获何梁何利基金“优秀志愿单位”荣誉称号。

2014年11月

北师大天文系翻译出版的《DK宇宙大百科》一书，荣获第八届吴大猷科学普及著作奖佳作奖，并入选科技部2016年全国优秀科普作品。

2014年11月4-8日

第十六届中国国际工业博览会在上海国家会议中心召开。国家授时中心“北斗高精度广域增强服务系统及用户终端”项目参加中国科学院团体展并获创新奖。

2014 年 12 月 15 日

“用于太阳磁场精确测量的中红外观测系统（AIMS）”获国家自然科学基金委批准。

2015 年

上海交通大学景益鹏教授当选为中国科学院数理学部院士。

2015 年 1 月

南极巡天望远镜 AST3-2 成功安装于南极冰穹 A。

2015 年 1 月 7 日

云南天文台与泰国国家天文研究所合作协议签字仪式在泰国首都曼谷举行。

2015 年 5 月 11 日

首届国际 GNSS（Global Navigation Satellite System）空间信号质量监测评估研讨会（GNSS Signal in space Quality Monitoring and Assessment（SMAS）Seminar）在中国西安曲江国际会议中心举行。

2015 年 5 月 22 日

北京天文馆被评为 2015-2019 年“全国科普教育基地”称号。

2015 年 6 月 8 日

国家航天局空间碎片监测与应用中心在国家天文台挂牌成立。

2015 年 6 月 23 日

紫金山天文台（紫金山园区）通过复核被续评为 2015-2019 年“全国科普教育基地”。

2015 年 8 月 6 日

500 米口径球面射电望远镜（FAST）工程反射面索网制造和安装工程通过竣工验收。

2015 年 8 月 17 日

紫金山天文台和海西州政府共建的德令哈天文科普馆正式开馆。

2015 年 8 月 30 日 -9 月 4 日

第六届海峡两岸天文望远镜与观测前沿技术研讨会在台湾花莲成功召开。

2015 年 9 月 10 日

中国天文学会开展“第五届黄授书奖”评选活动，清华大学王晓锋和南京大学程鑫获“第五届黄授书奖”。

2015 年 9 月 16 日

中山大学物理与天文学院正式成立，天文与空间科学研究院成为物理与天文学院的一部分。

2015 年 9 月 22-24 日

中国天文学会开展“第一届黄润乾奖”评选活动，上海交通大学景益鹏和北京大学吴学兵获

“第一届黄润乾奖”。

2015 年 9 月 29 日

暗物质粒子探测卫星征名活动启动仪式在紫金山天文台举行。根据本次征名活动的结果，该卫星后来被命名为“悟空”。

2015 年 10 月

新疆天文台申报的“射电天文与技术国际联合研究中心”获得科技部批准成立，并正式纳入国家级国际科技合作基地序列。

2015 年 10 月 19-21 日

中国天文学会 2015 年学术年会在北京大学召开。参会代表 800 多人，大会特邀报告 8 篇，分会场交流口头报告 300 多篇，张贴报告 30 多篇。

2015 年 11 月 22-25 日

第十六届东亚地区亚毫米波接收机技术研讨会在南京召开。

2015 年 11 月 26 日

斯洛文尼亚共和国总理米罗·采拉尔到访北京古观象台。

2015 年 12 月 1 日

新版《英汉天文学名词》在中国科协三峡科技出版资助计划支持下由中国科学技术出版社出版。

2015 年 12 月 14 日

“中国 SONG 望远镜”顺利通过现场验收。

2015 年 12 月 17 日

暗物质粒子探测卫星“悟空”号在酒泉卫星发射中心用“长征二号”丁运载火箭成功发射，至此我国终于实现了天文卫星零的突破。

2015 年 12 月 24 日

我国科学卫星系列首发星—暗物质粒子探测卫星“悟空”在升空后第 7 天，成功获取首批科学数据。

2016 年

自 2016 年起，“新疆青年科普行”系列活动已举办 40 多期天文、地理及其他学科科普讲座。

2016 年

云南天文台韩占文研究员获何梁何利奖。

2016 年

厦门大学天文学系获批天体物理与天文仪器福建省高校重点实验室。

2016 年 1 月

中国科学院各天文系统单位和国内部分高校青年天文工作者共同发起，向中国天文学会提出了设立青年工作委员会的建议。经中国天文学会理事会研究，批准设立中国天文学会青年工作委员会。

2016年1月1日

中国发射暗物质粒子探测卫星入选中国航天十大新闻和世界航天十大新闻。

2016年4月11日

紫金山天文台与中国科学技术大学签署共建科教融合天文与空间科学学院框架协议。

2016年4月12-15日

在南京主办了“第27届国际空间太赫兹技术研讨会（ISSTT）”，本次会议是ISSTT会议首次在中国召开。

2016年4月15日

澳大利亚总理马尔科姆·特恩布尔到访北京古观象台。

2016年5月

“中国地基大口径光学红外望远镜的科学与技术发展战略”香山科学会议在南京成功召开。

2016年6月

南京天文光学技术研究所成功完成乌兹别克斯坦MAO天文台1米望远镜改造。

2016年6月1-7日

FAST工程参展国家“十二五”科技创新成就展；6月3日，习近平、李克强、刘云山、王岐山等党和国家领导人先后前往FAST展台观看。

2016年6月7日

中美大学天文合作高峰论坛在北京大学举行，中美近20所顶尖大学校长参加。参加者应邀出席第七轮中美人文交流高层磋商全体会议，受到中国国务院副总理刘延东、美国国务卿约翰·克里的接见。

2016年7月20-22日

第五届“青年天文论坛”在山东威海举办。

2016年8月19日

云南天文台－香港天文学会宽视场测光巡天项目正式开始运行。

2016年9月15日

“天极”伽马暴偏振探测仪随“天宫二号”空间实验室发射升空。

2016年9月20日

中国天文学会开展“第十三届（2015-2016年度）张钰哲奖”评选活动，中科院国家天文台韩金林获“第十三届（2015-2016年度）张钰哲奖”。

2016年9月25日

FAST落成启用，中共中央总书记、国家主席、中央军委主席习近平发贺信。

2016年10月15日

时间频率与卫星导航学术研讨暨庆祝国家授时中心成立50周年会议在西安临潼举行。

2016年10月17日

在西安国家民用航天产业基地召开了国际GNSS监测评估系统国家授时中心分析中心资格审查会，审查组一致认为国家授时中心分析中心达到了iGMAS分析中心准入标准。

2016年11月1-3日

中国天文学会2016年学术年会在湖北武汉召开，参会代表750多人，大会特邀报告8篇，分会场交流口头报告337篇，张贴报告40篇。

2016年12月5-7日

“第二届中澳天体物理学研讨会暨中澳天文联合研究中心工作会议”（2nd Australia-China Workshop on Astrophysics）在江苏苏州举行。

2016年12月13日

国际上首例以无人值守工作模式运行的超宽带（0.75~15THz）太赫兹傅里叶光谱仪获得了南极冰穹A天文台址太赫兹至远红外谱段的大气透过率长周期实测数据。

2017年3月6日

紫金山天文台青岛观象台被评为2016年度全国优秀科普教育基地。

2017年3月28日

紫金山天文台入选首批中国十大科技旅游基地。（注：同期入选的还有FAST，国家授时中心）

2017年4月6日

上海交通大学物理与天文学院召开天文系成立大会。

2017年4月16-18日

中国天文学会开展“第二届黄润乾奖”评选活动，中国科学院国家天文台赵刚和南京大学戴子高获“第二届黄润乾奖”。

2017年5月4日

北京大学科维理天文与天体物理研究所举办学术讨论会庆祝成立十周年。

2017年5月10-12日

第六届“青年天文论坛”在贵州平塘县克度镇举办。

2017年5月15-19日

第一届中欧太阳物理会议在云南省昆明市成功举办。

2017 年 6 月 15 日

首颗X射线空间天文卫星“慧眼”号成功发射。

2017 年 6 月 16 日

国家空间科学中心组织召开先进天基太阳天文台（ASO-S）卫星工程立项综合论证报告评审会。ASO-S 卫星工程已具备立项条件，同意通过评审。

2017 年 6 月 21 日

中国天文学会开展“第六届黄授书奖”评选活动，北京大学东苏勃和中国科学技术大学王慧元获“第六届黄授书奖”。

2017 年 6 月 30 日

国家重大科研仪器研制项目（部门推荐）“太赫兹超导阵列成像系统”顺利通过国家基金委组织的项目验收。

2017 年 7 月 6-9 日

“第二届天文丝绸之路：中国与中亚的天文和天文考古交流国际学术研讨会”在乌鲁木齐召开。

2017 年 8 月

中国天文学会女天文工作者委员会正式成立。

2017 年 8 月

福建省天文学会联合全台高校天文社联盟办理“闽台大学生天文交流营”。

2017 年 8 月

闽台港澳四地天文同仁共聚一堂庆祝嘉义市天文协会成立三十周年。

2017 年 8 月 7-11 日

中国天文学会 2017 年学术年会在新疆乌鲁木齐隆重开幕。参会代表 800 多人，大会特邀报告 8 篇，分会场交流口头报告 359 篇，张贴报告 40 篇。

2017 年 8 月 18 日

中国南极巡天望远镜 AST3 合作团队利用正在中国南极昆仑站运行的望远镜 AST3-2 对 GW 170817 开展了有效的观测，并探测到此次引力波事件的光学信号。

2017 年 9 月

南京大学天文学科入选首批“双一流建设专业名单”。

2017 年 9 月

国家授时中心研究团队研制的光晶格冷原子锶（87）光钟（简称锶光钟）成功实现了闭环运行。

2017 年 9 月 3-6 日

第七届海峡两岸天文望远镜及仪器学术研讨会在伊春成功召开。

2017 年 9 月 6 日

由国家授时中心作为主要承研单位的转发式卫星导航试验系统第二阶段研制建设任务顺利通过中国卫星导航系统管理办公室组织的验收测试，定位测速授时（PVT）能力取得重大突破。

2017 年 9 月 28 日

“2017 年丝路天文科学研讨暨庆祝新疆天文台成立 60 周年会议”召开。

2017 年 10 月

被誉为下一代大视场自适应光学的多层共轭自适应光学试验系统在云南天文台 1 米新真空太阳望远镜（NVST）上成功闭环，在国内首次实时获得太阳活动区大视场高分辨力图像，成为继美国和德国世界上第三个掌握该技术的国家。

2017 年 10 月

南极巡天望远镜 AST3-2 成功实现无人值守越冬观测，为探测引力波源光学对应体做出贡献。

2017 年 10 月 10 日

由中国科学院主持发布 500 米口径球面射电望远镜（FAST）取得的首批成果，包括 6 颗新脉冲星。

2017 年 10 月 27 日

在上海组织完成“上海 65 米射电望远镜系统研制项目”总体验收。

2017 年 10 月 27 日

丽江日冕仪合作站揭牌仪式在丽江高美古天文观测站隆重举行。来自日本国立天文台、国家天文台与云南天文台的专家学者共同见证了挂牌仪式。

2017 年 11 月

云南天文台韩占文研究员当选为中国科学院院士。

2017 年 11 月

北师大出版的译著《宇宙图志》入选科技部 2017 年全国优秀科普作品。

2017 年 11 月 3 日

国家天文台－南非夸祖鲁－奈特大学（NAOC-UKZN）计算天体物理联合中心成立。

2017 年 11 月 16 日

新版英汉天文学词典正式上线，纳入中国虚拟天文台平台。

2017 年 11 月 24 日

紫金山天文台本部由鼓楼搬迁至仙林园区，张钰哲先生铜像揭幕仪式在仙林园区隆重举行。

2017 年 11 月 26 日 -12 月 9 日

海上丝绸之路国际青年天文学者培训班在昆明成功举办。

2017年11月27日

暗物质粒子探测卫星“悟空”首批科学成果新闻发布会在中国科学院举行。

2017年12月

组织FAST与费米LAT实验室签署了关于脉冲星研究的合作备忘录（FAST-Fermi_LAT合作备忘录）。

2017年12月

南京天文光学技术研究所成功研制Φ1.35米超薄非球面改正镜助美国ZTF项目初光。

2018年

暗物质粒子探测卫星—“悟空”号参与“庆祝改革开放40周年大型展览”。

2018年1月

中国科学院FAST重点实验室成立，挂靠国家天文台。

2018年2月

国家授时中心飞秒光梳及其应用研究小组成功研制国内首台光生超稳微波频率产生装置（简称光生超稳微波源），达到国际先进水平。

2018年2月

由南京天文光学技术研究所联合南京天文仪器有限公司为印度天体物理研究所（IIAP）研制的第二台太阳色球望远镜开始投入运行。

2018年2月22日

紫金山天文台近地天体望远镜新发现的一颗对地球构成潜在威胁的近地小行星（PHA），2018 DH1，于2018年3月27日18时18分（北京时间）在距离地球9.18个地月距离处飞掠地球。

2018年3月

中国科学技术大学与紫金山天文台共建大视场巡天望远镜项目启动。

2018年5月23日

国家授时中心申报的“十三五”国家重大科技基础设施“高精度地基授时系统”项目建议书获得了国家发展和改革委员会批复。

2018年6月21日

中山大学正式批准复办天文系。

2018年7月4日

先进天基太阳天文台（ASO-S）卫星工程启动会在北京召开。

2018年7月16-18日

中国科学院首届“率先杯”未来技术创新大赛在深圳市南山区紫荆山庄落下帷幕。国家授时中心“小型化高性能掺铒光纤飞秒光梳”项目荣获大赛优胜奖。

2018 年 8 月

北京天文馆朱进当选为 IAU 小天体命名委员会委员。

2018 年 8 月 6-8 日

中国天文学会开展“第三届黄润乾奖”评选活动，中国科学院云南天文台李焱获“第三届黄润乾奖”。

2018 年 8 月 14-17 日

第七届“青年天文论坛”在新疆乌鲁木齐举办。

2018 年 9 月

由南京天文光学技术研究所研制的 LAMOST 中分辨率光谱仪通过现场验收，LAMOST 开启二期中分辨率光谱巡天，该项目获得 2018 年度“十大天文科技进展”。

2018 年 9 月 13 日

国家重点研发计划“大科学装置前沿研究”重点专项“射电技术方法前沿研究”项目启动会在紫金山天文台召开。

2018 年 9 月 15 日

首届中国天泉湖天文论坛在盱眙开幕。

2018 年 9 月 27 日

中国天文学会开展“第十四届（2017-2018 年度）张钰哲奖”评选活动，紫金山天文台常进获“第十四届（2017-2018 年度）张钰哲奖”。

2018 年 10 月 27-31 日

中国天文学会 2018 年学术年会在昆明召开。10 月 31 日至 11 月 1 日，召开第十四次全国会员代表大会。参会代表 900 多人，大会特邀报告 9 篇，高级科普报告 2 场、专题讲座 2 场，分会场交流口头报告 388 篇，张贴报告 32 篇。

2018 年 10 月 27 日

召开首届中国天文学会女天文工作者交流会议。

2018 年 10 月 29 日

“极光计划”X 射线偏振测量立方星（PolarLight）、“天格计划”空间分布式伽马暴探测网（GRID）首颗实验星发射。

2018 年 10 月 30-31 日

中国天文学会第十四次全国会员代表大会在云南昆明召开，会议代表 128 人，选举产生了第十四届理事会，上海交通大学景益鹏任理事长。

2018 年 11 月

国家授时中心信号质量评估团队在陕西省洛南县建设的以 40 米天线为核心的 GNSS 空间信号

质量评估系统，实现了 GNSS 卫星导航信号（全球卫星导航系统空间信号）的高精度测试评估。

2018 年 11 月 5 日

联合国全球卫星导航系统国际委员会（ICG）第十三届大会在陕西西安开幕。

2018 年 11 月 6 日

紫金山天文台常进研究员荣获何梁何利基金“科学与技术进步奖”。

2019 年

“事件视界望远镜”项目发布了人类首张黑洞照片，上海天文台牵头组织国内团队参与。

2019 年

上海天文台建成的世界首台 SKA 区域中心原型机正式发布。

2019 年

厦门大学天文学系以会员单位身份加入中国天文学会。

2019 年

厦门大学天文学系与上海天文台签订新一轮共建深化合作协议。

2019 年 1 月 7 日

中国天文学会开展“第七届黄授书奖”评选活动，南京大学谢基伟和中国科学院紫金山天文台郭建华获“第七届黄授书奖”。

2019 年 2 月

中国“墨子号”量子科学实验卫星团队获美国科学促进会 2018 年度纽科姆·克利夫兰奖，以表彰该团队实现千公里级的星地双向量子纠缠分发。新疆天文台作为项目参与方获得一枚纽科姆·克利夫兰奖章。

2019 年 2 月 27 日

紫金山天文台领导的暗物质粒子探测卫星“悟空”号（DAMPE）科学团队研究成果“首次直接探测到电子宇宙射线能谱在 1TeV 附近的拐折”成功入选第 14 届“中国科学十大进展”。

2019 年 3 月 12 日

中国等七国正式签署平方公里阵 SKA 天文台公约。

2019 年 3 月 19 日

清华大学发布天文系成立的决定。

2019 年 3 月 25 日

经过中国天文学会十四届理事会第一次会议、常务理事会第一次会议批准和委员会内部严格筛选，信息委首届共计 42 名委员名单完整出炉。

2019年4月

中国科学技术大学发现首例双中子星并合形成磁星所驱动的X射线暂现源，获2019年度中国科学技术大学杰出研究校长奖，并入选2019年度中国十大天文科技进展。

2019年4月8日

中山大学天文系成为中国天文学会单位会员。

2019年4月12日

青海省科学技术厅、紫金山天文台、国家天文台签订青海省天文大科学装置台址遴选框架性合作协议。

2019年4月21日

清华大学天文系成立大会正式举行。

2019年6月3-5日

全国VLBI（甚长基线干涉测量技术）科学技术及应用研讨会在国家授时中心举行。

2019年6月5日

以国家天文台为依托单位，紫金山天文台、上海天文台、中科院计算机网络信息中心为共建单位申报的“国家天文科学数据中心”被正式列入国家科技资源共享服务平台名单。

2019年6月11日

由国家授时中心策划的“从指南针到北斗——中国古代导航展”在联合国维也纳国际中心环形大厅隆重开幕。

2019年6月14日

中国天文学会信息委成立大会暨信息化工作交流研讨会在南京召开，宣布信息委成立，同时为信息委官方网站揭幕：http://iwcc.china-vo.org/index.html。

2019年6月30日-7月5日

国际天文联合会第353号学术讨论会在上海召开，会议主题为“大型巡天观测时代的星系动力学研究”。

2019年7月

国家授时中心守时理论与方法研究室团队在采用自主研发的相干消色散数字终端，利用国家授时中心洛南昊平站40米口径射电望远镜获得毫秒脉冲星B1937+21轮廓的精细结构。

2019年8月5-9日

在云南丽江举办了“Ia型超新星前身星”国际学术会议。

2019年8月6-9日

第八届“青年天文论坛”在山东青岛举办。

2019年8月10日

青海省重大科技专项“天文大科学装置冷湖

台址监测与先导科学研究”项目启动会在青海省西宁市召开。

2019年9月2-5日

第八届海峡两岸天文望远镜与观测前沿技术研讨会在台湾高雄召开。

2019年9月6-10日

中国天文学会2019年学术年会在青海省德令哈召开，参会代表600多位，大会特邀报告11篇、专题讲座2场、高级科普报告2场、分会场交流报告253篇，社会推广宣传参展单位10多家，走进科技馆、走进校园科普报告2场。

2019年9月27日

国家重点基础研究发展计划（973计划）项目“110米大口径全可动射电望远镜关键技术研究”课题成功验收。

2019年9月28日

暗物质粒子探测卫星“悟空”第二批科学成果正式发布。

2019年10月

辽宁省唯一的万维望远镜互动式天文教室于凌源市第二高级中学落成，经信息委研究决定授予凌源市第二高级中学“信息委科普教育基地”。

2019年10月-2020年8月

第四届万维望远镜宇宙漫游创作大赛顺利举行。

2019年10月8-10日

中国天文学会开展“第四届黄润乾奖”评选活动，中国科学院高能物理研究所王建民获“第四届黄润乾奖”。

2019年10月14-21日

国际天文联合会主席，荷兰莱顿大学天体物理学教授艾文·范·迪舒克（Ewine van Dishoeck）教授访问云南天文台，10月21日访问紫金山天文台。

2019年10月14日-11月2日

第42届国际青年天文学家培训学校在昆明顺利举办。

2019年10月21日

第二届中国天泉湖天文论坛在江苏盱眙召开。

2019年10月22日

国家天文台、丽江市政府、云南天文台、云南大学四方签署合作框架协议，共同建设国家天文科学数据中心丽江分中心。

2019年10月29-31日

“首届江苏省中学天文科普教育研讨会”在南京大学成功召开。

2019 年 11 月

紫金山天文台常进研究员入选中国科学院数理学部院士。

2019 年 11 月 4-8 日

由上海交通大学、李政道研究所和上海天文台联合举办国际会议“星系与宇宙学上海合作会议”。

2019 年 11 月 18 日

紫金山天文台史生才研究员荣获何梁何利基金“科学与技术进步奖”。

2019 年 11 月 18 日

紫金山天文台和中国科学院行星科学重点实验室获评何梁何利基金“优秀志愿单位”。

2019 年 11 月 19 日

中国天文学会 2019 图书信息与期刊出版研讨会在国家授时中心召开。

2019 年 11 月 20 日

国内原创 4K 球幕影片《巡天先驱——郭守敬天文成就》通过国家天文科学数据中心向社会开放共享。

2019 年 11 月 24-26 日

第二十届东亚地区亚毫米波接收机技术研讨会在南京召开。会议由紫金山天文台、中国科学院射电天文重点实验室主办。

2019 年 11 月 27 日 -12 月 1 日

虚拟天文台与天文信息学 2019 年学术年会在百湖之城大庆举办。

2019 年 12 月

1.8 米太阳望远镜（Chinese Large Solar Telescope, CLST）实现初光并在调试中。该望远镜是国内首台 2 米级口径太阳望远镜。

2019 年 12 月

江苏省天文学会开展首届“江苏省天文学会科学技术奖”和“江苏省天文学会青年人才奖”评选活动。

2019 年 12 月 7 日

中山大学物理与天文学院天文系在中山大学珠海校区复办揭牌仪式。

2019 年 12 月 8-13 日

全球卫星导航系统国际委员会（International Committee on Global Navigation Satellite Systems）第十四届大会在印度班加罗尔召开。

2019 年 12 月 10 日

信息委成员完成的《基于虚拟天文台大数据的万维天象厅》项目荣获 2018 年度科技馆发展奖“展览奖”提名奖。

2019年12月16日

第一届全国天文光子学学术研讨会在南京天文光学技术研究所举行。

2019年12月21日

IAU100太阳系外行星世界命名中国（内地）活动命名结果媒体说明会在北京天文馆举行。

2019年12月26日

“天文科普公众开放日活动”于凌源市第二高级中学万维望远镜互动式天文教室成功举行。

2019年12月26-27日

第二届女天文工作者交流会在新疆天文台南山观测站顺利召开。

2019年12月27日

由国家授时中心和福建省泉州市人民政府联合建设的“海上丝绸之路时间中心”正式揭牌。

2019年12月30日

中国天文学会开展“第十五届（2019-2020年度）张钰哲奖”评选活动，北京大学吴月芳获“第十五届（2019-2020年度）张钰哲奖”。

2020年

北斗三号全球组网最终完成，其中的信息处理系统、星载氢原子钟和激光测距系统由上海天文台研发。

2020年

北京大学天文学科成立六十周年，为中国的天文教育和科学研究做出重要贡献。

2020年

厦门大学天文台项目通过验收。

2020年1月

中华人民共和国工业和信息化部正式公布第三批国家工业遗产名单，“国家授时中心蒲城长短波授时台”成功入选，成为中国科学院此次唯一入选的大科学装置。

2020年1月6-10日

由欧洲天文期刊《天文与天体物理》、EDP出版社、云南天文台联合主办的第5届青年天文工作者科技论文写作学校在云南天文台凤凰山本部顺利举办。

2020年1月9日

紫金山天文台成灼获评“典赞·2019科技江苏”年度科学传播人物。

2020年1月11日

FAST通过国家验收，正式开放运行。

2020 年 3 月 23 日

紫金山天文台发现的近地小行星 2020 FL2 于凌晨 04 时 38 分 24 秒在 0.38 地月距离（约 14.4 万公里）飞掠地球。这是目前近地天体望远镜发现的与地球轨道距离最近的近地小行星。

2020 年 5 月

北京大学科维理天文与天体物理研究所副所长沈雷歌研究员（Gregory J. Herczeg）被任命为国际三十米望远镜项目（Thirty Meter Telescope,TMT）科学咨询委员会主席和美国天文学会期刊（AAS Journals）科学编辑。

2020 年 6 月 21 日

“夏至未至，十年归期”公众天文科普活动于凌源市万人广场成功举办，共赏日偏食。

2020 年 6 月 21 日

我国南方可见日环食，中科院天文科普联盟在多地组织举办日环食观测与网络直播活动。

2020 年 7 月 25 日

由南京大学天文与空间科学学院领衔研发的“龙虾眼 X 射线探测卫星”搭载长征四号乙运载火箭，在太原卫星发射中心成功发射入轨。

2020 年 7 月 26 日

华中科技大学天文学系成立大会正式举行。

2020 年 7 月 29 日

天文学名词委组织力量首次将国际天文学联合会（IAU）公布的全部 1950 条火星地形地貌名称全部译为中文，并向全社会公开发布。包含第一批 814 条地形地貌名称译名（7 月 29 日公布），第二批 1136 条火星陨击坑译名（8 月 24 日公布）。

2020 年 8 月

南京大学 1 米时域巡天望远镜在南京天文光学技术研究所顺利通过出所验收。

2020 年 9 月 19 日

以中山大学为牵头单位的国家级平台——中国空间站工程巡天望远镜（CSST）粤港澳大湾区科学中心在学院举行了揭牌仪式，为天文人才培养提供了国内一流且具有国际竞争力的科研平台。

2020 年 9 月 29 日

“景东 120m 脉冲星射电望远镜（JRT）研制”重大科技项目启动仪式在 JRT 台址——云南省普洱市景东县太忠镇徐家坝举行。

2020 年 10 月 12-14 日

“中国天文学会 2020 年学术年会”召开。首次采取线上线下相结合的方式，参会代表 1000 多人，大会特邀报告 12 篇、专题讲座 2 场、高级科普报告 1 场，分会场交流报告 251 篇。

2020 年 11 月

北师大天文系利用我国 FAST “天眼”射电望远镜和“慧眼”X 射线望远镜认证并研究了快速射电暴的磁星起源，被 *Nature* 和 *Science* 评选为 2020 年度十大科学发现。

2020 年 11 月

时频领域国际顶级期刊 *Metrologia* 发表了国家授时中心时间频率基准实验室 GNSS 时间性能评估团队的学术论文“Analysis on the time transferperformance of BDS-3 signals”（“北斗三号”信号的时间传递性能分析）。研究表明，“北斗三号”的时间传递性能相对于“北斗二号”提高了 50% 以上。

2020 年 11 月 21 日

中国空间站工程巡天望远镜北京大学科学中心揭牌成立。

2020 年 11 月 25-29 日

虚拟天文台与天文信息学 2020 年学术年会在鹭岛厦门举办。

2020 年 12 月

SKA 专项宇宙黎明和再电离探测立项。

2020 年 12 月 6-8 日

江苏省天文学会 2020 年学术年会暨成立 90 周年庆祝大会在南京召开。

2020 年 12 月 15 日

中国天眼 FAST 望远镜在快速射电暴方面的研究成果入选《自然》2020 年十大科学发现。

2020 年 12 月 19 日

“嫦娥五号”任务月球样品接收活动在国家天文台举行。

2021 年

国家授时中心《时间的真相》获科技部、中科院、安徽省优秀科普作品奖。

2021 年 1 月

江苏省天文学会“仰望星空，探秘天文新视野”科普服务项目获得中国科协“2020 年度科技志愿服务项目先进典型”。

2021 年 1 月 11 日 -2 月 1 日

《走进天文软件的“无冕之王”天文软件大师新年讲座系列》成功举行。

2021 年 1 月 25 日

中国天文学会开展“第八届黄授书奖”评选活动，北京大学田晖和国家天文台闫宏亮获“第八届黄授书奖”。

2021 年 2 月

据国际权度局（BIPM）官方数据统计，2020 年，由国家授时中心保持的我国时间基准UTC（NTSC）性能继续位居世界前列，国家授时中心为国际原子时贡献权重7.0%，世界第三。

2021年2月5日

习近平总书记亲切会见了“中国天眼”项目负责人和科研骨干。

2021年2月22日

习近平总书记会见探月工程“嫦娥五号”任务参研参试人员代表（新疆天文台台长王娜研究员作为“嫦娥五号”VLBI测轨分系统南山站参研参试人员代表）。

2021年3月

国家天文科学数据中心分中心共建签约授牌仪式在北京国家天文台举行。天津大学、广州大学与国家天文科学数据中心签订合作共建协议，将分别承担国家天文科学数据中心技术研发创新中心和国家天文科学数据中心粤港澳大湾区分中心的建设任务。

2021年3月31日

中国天眼FAST正式对全球开放。

2021年5月

江苏省天文学会获得江苏省民政厅“中国社会组织评估等级4A等级”，这是学会历史上获评的最高等级。

2021年5月7-9日

中国天文学会开展“第五届黄润乾奖”评选活动，中国科学技术大学王挺贵和北京师范大学朱宗宏获“第五届黄润乾奖”。

2021年5月11日

中国科学技术大学–紫金山天文台2.5米大视场巡天望远镜项目（Wide Field Survey Telescope，简称WFST）开工奠基仪式在青海省海西州茫崖市冷湖镇赛什腾山天文观测台址隆重举行。

2021年5月19日

暗物质粒子探测卫星“悟空”第三批科学成果新闻发布会在紫金山天文台举行。

2021年5月21日

新疆天文台微波技术实验室被正式认定为新疆微波技术重点实验室（自治区级重点实验室）。

2021年6月9日

中山大学与国家天文台签订建设“中国天眼联合研究中心”合作协议。

2021年6月26日

紫金山天文台多应用巡天望远镜阵项目在青海冷湖开工。

2021 年 7 月

国家授时中心时间频率测量与控制研究室团队成员经过多年研究，在时钟控制、远程时间比对等技术方面取得了一系列成果，完成了标准时间远程复现系统工程建设，通过了验收测试，具备了为国内外用户提供与国家标准时间偏差优于5ns 的时间信号的能力。

2021 年 7 月

全球最大的天文馆——上海天文馆正式对外开放。馆内建成全国第一座专门反映天文数字化研究及数据可视化应用的教育空间“天文数字实验室”。

2021 年 8 月

福建省天文学会建成永泰县盖洋乡“乌石庄园”星空研学观测基地。

2021 年 9 月 7 日

国家空间科学中心国家空间科学数据中心、紫金山天文台联合公开发布暗物质粒子探测卫星“悟空”首批伽马光子科学数据。

2021 年 9 月 25 日

国家重大科技基础设施项目高精度地基授时系统拉萨授时监测站建设工程启动仪式在拉萨市经开区举行。

2021 年 9 月 25-28 日

第九届海峡两岸天文望远镜及仪器学术研讨会在贵州平塘召开。

2021 年 10 月

第五届万维望远镜宇宙漫游创作大赛正式启动，计划于 2022 年 7 月收官。本届大赛将在获奖作品中推选 6 部作品参加首届国际宇宙漫游创作大赛。

2021 年 10 月 14 日

南京大学参与研发的我国首颗太阳探测科学技术试验卫星“羲和号”在太原卫星发射中心由“长征二号”丁运载火箭成功发射升空，开启我国的空间探日时代。

2021 年 10 月 14 日

2021 年全国时间频率学术会议在甘肃省敦煌市举行。

2021 年 11 月

武汉大学与国家授时中心等联合申报的“北斗 /GNSS 全球连续监测评估数据中心系统关键技术及应用”项目获中国卫星导航定位协会颁发的卫星导航定位科技进步奖一等奖。

2021 年 11 月

国家重大科研仪器研制项目“2.5 米大视场高

分辨率太阳望远镜”批准立项。

2021 年 11 月 18 日

紫金山天文台史生才研究员当选中国科学院数学物理学部院士。

2021 年 12 月

国际天文学联合会天文教育办公室（IAU OAE）中国中心在北京天文馆设立。

2021 年 12 月

国家天文科学数据中心与华中师范大学、河北师范大学、西华师范大学签订合作协议，共建教育研发应用中心，协同实施基于真实科学数据的课程研发、教育教学实践，实现科学数据、信息技术与教育教学内容深度融合。

2021 年 12 月 2 日 -6 日

中国天文学会 2021 年学术年会在四川南充召开。参会代表 1000 多人，采取线上线下相结合的方式，大会特邀报告 13 篇、专题讲座 2 场、高级科普报告 2 场、9 个分会场、交流报告 379 篇。

2021 年 12 月 10 日

紫金山天文台联合南京地质古生物研究所合作研究“嫦娥五号”首批月球样品取得重要进展。表明“嫦娥五号”着陆区历史上可能曾经发生过多次火山喷发活动，将有望解读月幔源区不同物质成分、火山岩浆形成的能量来源和月球晚期火山活动的精细时空分布规律。

2021 年 12 月 16 日

中国大型光学望远镜前沿技术学术研讨会在南京举办。

2022 年

佘山天文台历史上最大规模的升级改造在松江区人民政府、上海市科委和上海天文台共同支持下完成。

2022 年 1 月 29 日

紫金山天文台译著《宇宙全书：国家地理新视觉指南》荣获 2020 年度全国优秀科普作品。

2022 年 2 月 10 日

中国天文学会开展“第十六届（2021–2022 年度）张钰哲奖”评选活动，中国科学院国家天文台赵永恒获“第十六届（2021–2022 年度）张钰哲奖”。

2022 年 2 月 16 日

著名天文学家、中国现代天文学重要奠基人、中国科学院院士张钰哲先生诞辰 120 周年，“星耀中华 风范千秋”弘扬科学家精神暨纪念张钰哲先生诞辰 120 周年活动在南京举办。

2022 年 3 月 30 日

中国科协命名“2021–2025 年第一批全国科普教育基地”，紫金山天文台的紫金山园区、青岛观象台、青海观测站三个基地同时入选。

2012–2022 年获得国家一、二等奖和特等奖的天文项目（含国家级个人奖项）

年份	国家奖类别	获奖名称（主要完成人）
2012	国家科学技术进步特等奖	“嫦娥二号”工程（国家天文台）
2012	国家自然科学奖二等奖	“高能电子宇宙射线能谱超出”的发现（常进等）
2013	国家自然科学奖二等奖	大样本恒星演化与特殊恒星的形成（韩占文等）
2014	第 11 届中国青年女科学家奖	陈雪飞
2014	国家科学技术进步二等奖	星地融合广域高精度位置服务关键技术（国家天文台）
2014	全国优秀科技工作者十佳提名奖	邓晓华（南昌大学）
2014	全国优秀科技工作者	伍歆（南昌大学）
2015	全国先进工作者	王娜
2016	国家科学技术进步特等奖	“北斗二号”卫星工程（国家授时中心、上海天文台）
2016	国家科学技术进步一等奖	“嫦娥三号”工程（国家天文台）
2017	全国创新争先奖章	南仁东
2017	“时代楷模”荣誉称号	南仁东
2019	第 16 届中国青年女科学家奖	李婧
2019	国家科学技术进步二等奖	“北斗”性能提升与广域分米星基增强技术及应用（陈俊平、曹月玲、巩秀强）

年份	国家奖类别	获奖名称（主要完成人）
2020	国家自然科学奖二等奖	基于高精度脉泽天体测量的银河系旋臂结构研究（徐烨、郑兴武、张波等）
2020	第 25 届“中国青年五四奖章集体”	紫金山天文台“悟空”号卫星载荷与科学团队
2021	国家科学技术进步特等奖	“嫦娥四号”工程（国家天文台）
2021	第 17 届中国青年女科学家团队奖	南京天文光学技术研究所袁祥岩研究员负责的“南极巡天望远镜研制团队”

2012–2022 年获得省部级一等奖和特等奖的天文项目（含省部级个人奖项）

年份	省部级奖类别	获奖名称（主要完成人）
2011	云南省科学技术（自然科学）一等奖	双星演化与特殊恒星形成（陈雪飞等，该奖项于 2012 年颁发）
2012	江西省自然科学一等奖	天体系统的非线性结构（刘三秋、伍歆、邓新发）
2013	军队科学技术进步一等奖	“北斗二号”试验卫星在轨技术试验（杨旭海、雷辉）
2014	云南省科学技术（自然科学）一等奖	共双星系外行星和小质量天体的观测与研究（钱声帮等）
2014	上海市科技进步奖一等奖	“嫦娥三号”和“玉兔”软着陆的 VLBI 实时精密测定轨和月面定位（上海天文台）
2015	云南省科学技术特等奖	一米新真空太阳望远镜研制及其在太阳观测中的应用（刘忠等）

年份	省部级奖类别	获奖名称（主要完成人）
2015	上海市自然科学一等奖	探索暗物质晕中的星系形成和演化（杨小虎）
2015	广西省科学技术一等奖	伽马射线暴及其余辉的辐射成分和物理起源（广西大学、国家天文台）
2015	教育部科技进步奖一等奖	光学和近红外太阳爆发探测望远镜（方成等）
2015	国防科学技术进步特等奖	“嫦娥三号”工程（国家天文台）
2015	上海市科学技术一等奖	探索暗物质晕中的星系形成与演化（上海天文台）
2016	广西省科学技术一等奖	500MPa 应力幅耐疲劳高精度索网关键技术的研究与应用（国家天文台）
2016	教育部第六届高等学校教学名师奖	向守平
2016	中国科学院教育教学成果一等奖	科教结合英才班－物理创新研究型人才培养模式探索和实践（中科大天文系）
2017	北京市科学技术一等奖	500 米口径球面射电望远镜超大空间结构工程创新与实践（南仁东等）
2017	中国科学院杰出科技成就奖	500 米口径球面射电望远镜 (FAST) 工程研究集体（南仁东等）
2017	新疆维吾尔自治区科技进步一等奖	脉冲星转动不稳定性的观测与研究（新疆天文台）
2017	教育部自然科学优秀成果一等奖	早期宇宙中最亮的类星体中的最大质量黑洞（吴学兵等）
2017	军队科学技术进步一等奖	空间目标白天高分辨力光电成像关键技术及应用（国家天文台）
2018	上海市科技进步特等奖	上海 65 米射电望远镜系统研制（上海天文台）
2018	辽宁省科学技术进步一等奖	500 米口径射电望远镜柔性并联索驱动系统技术及装备（朱文白等）
2018	江苏省科学技术奖一等奖	“悟空”号暗物质粒子探测器（常进等）
2019	江苏省科学技术奖一等奖	LAMOST 的核心创新和关键技术（苏定强等）
2019	中国科学院杰出科技成就奖	LAMOST 工程研究集体（苏定强等）

年份	省部级奖类别	获奖名称（主要完成人）
2019	贵州省科学技术进步一等奖	FAST 开挖系统关键技术应用研究及推广（朱博勤等）
2019	安徽省科学技术进步一等奖	LAMOST 焦面光纤定位装置（赵永恒等）
2019	安徽省教学成果特等奖	探索科教融合培养天文精英人才的新模式（袁业飞等）
2019	贵州省科学技术进步一等奖	FAST 主动反射面技术创新与实践（姜鹏等）
2019	云南省科技进步（自然科学）特等奖	太阳爆发过程中的电磁相互作用（林隽等）
2019	国防科学技术进步特等奖	“嫦娥四号”工程（李春来等）
2019	上海市科学技术一等奖	被动型氢原子钟（上海天文台）
2019	军队科学技术进步一等奖	小型光抽运铯束原子频率标准（张首刚等）
2019	新疆青年五四奖章	新疆天文台探月工程团队
2020	河北省科学技术进步一等奖	液压元件 / 系统及系统群可靠性增长关键技术及应用（朱明等）
2020	教育部“万人计划”教学名师	李向东
2021	中国科学院杰出科技成就奖	月球与深空探测地面应用系统研究集体（李春来等）
2021	教育部课程思政教学名师和团队	李向东等
2021	江苏省教学成果奖特等奖	从《天文探秘》到《宇宙简史》理科课程思政的探索与实践（李向东等）
2021	云南省科学技术（自然科学）一等奖	太阳活动能量传输和释放的精细物理过程（闫晓理等）
2021	中国科学院先进工作者	张首刚
2021	江苏省“最美科技工作者”	杨戟
2021	2020 年中科院优秀科普图书	《时间的真相》（李孝辉）

中国天文机构

CHINESE ASTRONOMICAL INSTITUTIONS

（1）中国科学院下属单位（按天文台、研究院所、观测站，及其成立的时间顺序）

中国科学院紫金山天文台

Purple Mountain Observatory, Chinese Academy of Sciences

地址：江苏省南京市元化路10号，邮政编码：210023 (Address：10 Yuanhua Rd.，Nanjing 210023，Jiangsu Province, P.R.China),
Tel：025-83332000，Fax：025-83332091，
Email：pmoo@pmo.ac.cn
http://www.pmo.cas.cn/

中国科学院国家天文台

National Astronomical Observatories, Chinese Academy of Sciences

地址：北京市朝阳区大屯路甲20号，邮政编码：100101（Address：20A Datun Road，Chaoyang District，Beijing 100101，P.R.China），Tel：010-64888732，Email：goffice@nao.cas.cn，http://www.nao.cas.cn/

中国科学院上海天文台

Shanghai Astronomical Observatory, Chinese Academy of Sciences

地址：上海市南丹路80号，邮政编码：200030 (Address：80 Nandan Rd.，Shanghai 200030，P.R.China), Tel：021-64386191，Fax：021-64384618，http://www.shao.cas.cn/

中国科学院国家授时中心

National Time Service Center, Chinese Academy of

Sciences

地址：西安市临潼区书院东路3号，邮政编码：710600 (Address：P.O.Box，Lintong，Xi'an 710600，P.R.China),Tel：029-83890326，Fax：029-83890196, http://www.ntsc.cas.cn/

中国科学院云南天文台

Yunnan Astronomical Observatory, Chinese Academy of Sciences

地址：云南省昆明市官渡区羊方旺396号，邮政编码：650216 (Address：396 Yangfangwang, Guandu District, Kunming 650216，Yunnan，P.R.China)，Tel：0871-3920919，Fax：0871- 3920599，http://www.ynao.cas.cn/

中国科学院新疆天文台

Xinjiang Astronomical Observatory, Chinese Academy of Sciences

地址：新疆乌鲁木齐市新市区科学一街150号，邮政编码：830011 (Address：150 Kexue 1 Rd，Urumqi，Xinjiang 830011，P.R.China)，Tel：0991-3689007，3838007，Fax：0991-3838628，Email：uao@uao.ac.cn, http://www.xao.cas.cn/

中国科学院自然科学史研究所

The Institute for the History of Natural Science, Chinese Academy of Sciences

地址：北京市海淀区中关村东路55号，邮政编码：100190 (Address：55 Zhongguancun East Rd, Haidian District，Beijing l00190，P.R. China)，Tel：010-57552515，Fax：010-57552567，webmaster@ihns.ac.cn，http：//www.ihns.ac.cn/

中科院南京天文仪器有限公司

CAS Nanjing Astronomical Instruments Co, Ltd

地址：江苏省南京市花园路6-10号， 邮政编码：210042 （Address：6-10 Huayuan Rd，
Nanjing 210042, Jiangsu Province, P.R. China),
Tel：025-85411830,
Email：sales@nairc.ac.cn, http://www.nairc.com

中国科学院光电技术研究所

Institute of Optics and Electronic, Chinese Academy of Sciences

地址：中国四川省成都市双流350信箱，邮政编码：610209 （Address：Box 350, Shuangliu, Chengdu City 610209, Sichuan Province，P.R.China），Tel: 028-85100341， Fax: 028-85100268， http://www.ioe.ac.cn/

中国科学院高能物理研究所粒子天体物理重点实验室

Key Laboratory of Particle Astrophysics Institute of High Energy Physics, Chinese Academy of Sciences

地址：北京市石景山区玉泉路19号乙，邮政编码：100049 (Address：19 Yuquanlu Road，Shijingshan District，Beijing 100049，P.R.China)，Tel：010-88236208，传真：010-88233260，

http://www.ihep.cas.cn/zdsys/lzttlab

中国科学院国家天文台南京天文光学技术研究所

Nanjing Institute of Astronomical Optics and Technology, National Astronomical Observatories, Chinese Academy of Sciences

地址：江苏省南京市太平门外板仓街188号，邮政编码：210042 (Address：188 Bancang Rd，Nanjing 210042，P.R. China), Tel: 025-85430617, Email: lhxie@niaot.ac.cn, http://www.niaot.cas.cn/

中国科学院大学人文学院

School of Humanities, University of Chinese Academy of Sciences

地址：北京市石景山区玉泉路19号（甲），邮政编码：100049 (Address: 19 Yuquan Rd, Shijingshan District, Beijing 100049, P.R.China), Tel: 010-88256005, https://renwen.ucas.ac.cn/

中国科学院大学天文与空间科学学院

School of Astronomy and Space Science, University of Chinese Academy of Sciences

地址：北京市石景山区玉泉路19号（甲），邮政编码：100049 (Address: 19 Yuquan Rd, Shijingshan District, Beijing 100049, P.R.China), 010-64807969, https://astro.ucas.ac.cn/

中国科学院精密测量科学与技术创新研究院

Innovation Academy for Precision Measurement Science and Technology, CAS

地址：湖北省武汉市徐东大街340号，邮政编码：430077 (Address：340 Xudong Rd, Wuhan，Hubei 430077，P.R.China)，Tel：027-68881355，Fax：027-68881362，http://www.apm.ac.cn/

中国科学院紫金山天文台青岛观象台

Qingdao Observatory, Purple Mountain Observatory, Chinese Academy of Sciences

地址：山东省青岛市市南区观象二路21号，邮政编码：266003 (Address：21 Guanxiang 2 Rd，Shinan District Qingdao City 266003, Shandong Province ,P.R.China),
Tel：0532-82828405，Email：qdgxt@pmo.ac.cn，http://qdgxt.kepu.net.cn/

中国科学院国家天文台长春人造卫星观测站

Changchun Observatory, National Astronomical Observatories, Chinese Academy of Sciences

地址：吉林省长春市净月潭西山，邮政编码：130117 (Address：Jingyuetan，Changchun，Jilin 130117，P.R.China), Tel：0431-84513834，Fax：0431-84516073, Email：webmaster@cho.ac.cn, http://www.cho.cas.cn/

（2）高等院校及所属研究单位（按天文类、物理类，及其成立的时间顺序）

南京大学天文与空间科学学院

School of Astronomy & Space Science, Nanjing University

地址: 江苏省南京市栖霞区仙林大道163号天文楼，邮政编码：210023 (Address：TianWen Building, No. 163, Xianlin Avenue, Qixia District，Nanjing 210023，P.R.China)，Tel：025-83592882，Fax：025-83235192，http://astronomy.nju.edu.cn/

北京师范大学天文系

Department of Astronomy, Beijing Normal University

地址：北京新街口外大街19号，邮政编码：100875 (Address：19 Xinjiekouwai Rd，Beijing 100875，P.R.China)，Tel：010-62207832，Fax：010-62206319，http://www.bnu.edu.cn/

中国科技大学天文学系

Department of Astronomy, University of Science and Technology of China

地址：安徽省合肥市金寨路96号，邮政编码：230026 (Address：96 Jinzhailu Road，Hefei，Anhui 230026，P.R.China)，Tel：0551-3601861，Fax：0551-3601853，Email：cfaoffice@ustc.edu.cn, http://astro.ustc.edu.cn/

北京大学物理学院天文系

Department of Astronomy, School of Physics,Peking University

地址：北京海淀颐和园路5号，邮政编码：100871 (Address：5 Yiheyuan Rd，Haidian District，Beijing 100871，P.R.China)，Tel：010-62751134，Fax：010-62765031，http://vega.bac.pku.edu.cn/

北京大学科维理天文与天体物理研究所

The Kavli Institute for Astronomy and Astrophysics (KIAA)

地址：北京海淀颐和园路5号，邮政编码：100871 (Address：5 Yiheyuan Rd，Haidian District，Beijing 100871，P.R.China)，Tel：010-62756692，Fax：010-62767900，https://kiaa.pku.edu.cn/

中科院国家天文台－贵州大学天文联合中心

Astronomical United Center of National Astronomical Observatories and Guizhou University

地址: 贵州省贵阳市花溪区 邮编: 550025（Address: Huaxi district of Guiyang City, Guizhou Province，550025, P.R.China)，Tel：0851-8292178，Fax：0851-3621956，Email：po@gzu.edu.cn, http://www.gzu.edu.cn/

山东大学（威海）空间科学与物理学院

School of Space Science and Physics Shandong University at Weihai

地址：山东省威海市文化西路180号，邮政编码：

264209 (Address：180 WenHua West Rd，Weihai 264209，P.R.China), http://apd.wh.sdu.edu.cn/

广西大学－国家天文台天体物理和空间科学研究中心

GXU-NAOC Center for Astrophysics and Space Sciences

广西省南宁市大学路100号广西大学物理学院，邮政编码:530004（Address：100 Daxue Rd，Nanning 530004，Guangxi Province, P.R.China), Tel: 771-3237386, http://astro.gxu.edu.cn/

厦门大学天文学系

Department of Astronomy, Xiamen University

地址：福建省厦门市思明区曾厝垵西路1号，邮政编码：361005 (Address：No. 1, Zengcuo'an West Road, Siming District, Xiamen City, Fujian Province 361005, P.R.China), Tel:0592-2186393, Fax：0592-2189426, Email: astro@xmu.edu.cn, http://astro.xmu.edu.cn

河北师范大学物理学院空间科学与天文系

College of Physics, Hebei Normal University

地址：河北省石家庄市南二环东路20号，邮政编码：050024（Address: No.20 Road East. 2nd Ring South, Yuhua District, Shijiazhuang, Hebei, 050024, P.R. China)，Tel: 0311-80789777，
Email: wlxy@hebtu.edu.cn, http://www.hebtu.edu.cn/

云南大学中国西南天文研究所、天文系

South-Western Institute For Astronomy Research、Department of Astronomy, Yunnan University

地址：云南省昆明市呈贡区云南大学呈贡校区天文楼，邮政编码：650500（Address：Astronomy Building, Yunnan University，Chenggong District，Kunming City 650500，Yunnan Province, P. R. China)，Tel: 0871-65931962，Fax: 0871-65931691，Email: swifar@ynu.edu.cn 或者 zjjiang@ynu.edu.cn，http://www.swifar.ynu.edu.cn/ 或者 http://www.astro.ynu.edu.cn/

中山大学物理与天文学院天文系

Department of Astronomy, School of Physics and Astronomy, Sun Yat-sen University

地址：中国广东省珠海市香洲区唐家湾大学路2号，邮政编码：519082 (Address：2 Tangjiawandaxue Rd., Xiangzhou District, Zhuhai City 519082, Guangdong Province, P.R.China), Tel: 0756-3668932，http://spa.sysu.edu.cn/

西华师范大学物理与天文学院

School of Physics and Astronomy, China West Normal University

地址：四川省南充市师大路1号，邮政编码：637002 (Address：1 Shida Rd. Nanchong City 637002, Sichuan Province, P.R.China)，Tel: 0817-2568348，https://pei.cwnu.edu.cn/index.htm

上海交通大学天文系

Department of Astronomy, Shanghai Jiao tong University

地址：上海市东川路 800 号理科楼群 5 号楼 5 层，邮政编码：200240 (Address：5th Floor, No. 5 Science Building, 800 Dongchuan Rd.，Shanghai 200240，P.R.China), Tel：021–54740262，021–54740257，021–54740240，Fax:021–54741040，http://astro.sjtu.edu.cn/zh/

山东大学空间科学研究院

Institute of Space Sciences, Shandong University

地址：山东省威海市文化西路 180 号，邮政编码：264209 (Address：180 WenHua West Rd，Weihai, Shandong, 264209，P.R.China), Tel: +86 (0631) 56788881, http://iss.wh.sdu.edu.cn/

贵州师范大学物理与电子科学学院天文系

Department of Astronomy, School of Physics and Electronic Science, Guizhou Normal University

地址：贵州省贵阳市宝山北路 116 号，电话：0851–83227328，https://phy.gznu.edu.cn/

清华大学天文系

Department of Astronomy, Tsinghua University

地址：北京海淀清华园 1 号，邮政编码：100084 (Address：1 Qinghuayuan，Haidian District，Beijing 100084，P.R.China)，Tel：010–62794613，Fax：010–62794613，http://www.tsinghua.edu.cn/

华中科技大学天文学系

Department of Astronomy, Huazhong University of Science and Technology

地址：湖北省武汉市珞瑜路 1037 号，邮政编码：430074（Address：1037 Luoyu Rd., Wuhan City 430074，Hubei Province, P.R.China)，Email：astro@hust.edu.cn，http://astro.hust.edu.cn/

广州大学物理与材料科学学院天文系

Department of Astronomy, School of Physics and Materials Science, Guangzhou University

地址：广东省广州市大学城外环西路 230 号，邮政编码：510006（Address：230 Waihuan West Rd，University Town, Guangzhou 510006，Guangdong Province, P.R.China), Tel：020–39366871, Email：webmaster@gzhu.edu.cn, http://spee.gzhu.edu.cn

泰山学院天文台

Taishan University Observatory

山东省泰安市东岳大街 525 号泰山学院，邮政编　码：271021（Address：525 Dongyue Rd，Taian，271021，Shandong Province，P.R.China），Tel:538–6715613, https://www.tsu.edu.cn/

南昌大学物理系

Department of Physics, Nanchang University

地址：江西省南昌市红谷滩新区学府大道 999 号，邮政编码：330031 (Address：999 Xuefu Rd，Honggutan new District ,Nanchang 330031, P.R.China),

Tel:：0791-3969099, Fax: 0791-3969069，http://www.ncu.edu.cn/

南京师范大学物理科学与技术学院

School of Physics and Technology,Nanjing Normal University

地址：南京市仙林大学城文苑路1号，邮政编码：210023（Address：1 Wenyuan Rd，Xianlin ,Nanjing 210023，P.R. China), Tel：025-85898966，Fax：025-85898223，http://physics.njnu.edu.cn/

华中师范大学物理科学与技术学院

College of Physics Science and Technology, Central China Normal University

地址：湖北省武汉市珞瑜路152号，邮政编码：430079，（Address：152 Luoyu Rd，Wuhan 430079，P.R. China), http://www.ccnu.edu.cn/

武汉大学测绘学院

School of Geodesy and Geomatics, Wuhan University

地址：湖北省武汉市珞瑜路129号，邮政编码：430079(Address: 129 Luoyu Rd,Wuhan 430079,P.R. China), Tel：027-87886686，Fax：027-68778371，http://main.sgg.whu.edu.cn

上海师范大学数理学院物理系

Department of Physics, Mathematics & Science College, Shanghai Normal University

地址：上海市桂林路100号，邮政编码：200234 (Address：100 Guilin Rd，Shanghai 200234，P.R.China), Tel:021-64322000, http://mathsc.shnu.edu.cn/

黔南民族师范学院物理与电子科学系

School of Physics and Electronics, Qiannan Normal University for Nationalities

地址：贵州省都匀市，电话：0854-8738063, 邮政编码：558000, https://wlx.sgmtu.edu.cn/

山东大学前沿交叉科学青岛研究院

Institute of Frontier and Interdisciplinary Sciences, Shandong University

地址：山东省青岛市即墨滨海路72号华岗苑东楼324（Address: 324 Huagangyuan, 72 Binhai Rd, Jimo, Qingdao, Shandong），Tel：（0532）58630258, E-mail：qyy@sdu.edu.cn, http://www.frontier.qd.sdu.edu.cn/

德州学院－物理与电子信息学院

College of Physics and Electronic Information, Dezhou University

地址：山东省德州市德城区大学西路566号，邮政编码：253023（Address：No. 566 University Rd. West, Decheng District, Dezhou City, Shandong Province），TEL:0534-8985888; http://www.dzu.edu.cn/

曲阜师范大学物理工程学院

School of Physics and Physical Engineering, Qufu Normal University

地址：山东省济宁市曲阜市静轩西路57号，邮政编码：273165 (Address：57 Jingxuan West Rd. Qufu City, Jining 273165, Shandong Province, P.R.China), Tel: 0537–4456092，https://physics.qfnu.edu.cn/

（3）其他天文机构（按建立的时间顺序）

北京天文馆

Beijing Planetarium

地址：北京市西直门外大街138号，邮政编码：100044 (Address：138 Xizhimenwai Road，Beijing 100044，P.R. China)，Tel：010–51583311，Fax：010–51583312，bjtwg@126.com, http://www.bjp.org.cn/

苏州市青少年天文观测站

Suzhou Astronomy Observatory

地址：苏州市姑苏区人民路979号，邮政编码：215000 (Address：979 Renmin Rd., Gusu District, Suzhou City 215000, P.R.China), Tel：0512–65227765, Email: suzhoutianwen@126.com, http://www.szao.org.cn/

上海天文馆

Shanghai Astronomy Museum

地址：上海市浦东新区临港大道380号，邮政编码：201306（Address：380 Lingang Rd., Pudong New District, Shanghai 201306, P.R. China）, Tel: 021–50908563, https://www.sstm–sam.org.cn/

ENGLISH VERSION

A Decade of Chinese Astronomy to Ahead

Jing Yipeng 14th President of Chinese Astronomical Society

Since the 18th National Congress of the Communist Party of China, Chinese astronomers have been forging ahead in various fields of astronomy. They have made remarkable achievements in the construction of scientific research team, the development of astronomical equipment, the publications, and their indicators in some aspects have ranked among the top in the world. 2021 coincides with the centenary of the founding of the Communist Party of China, and 2022 is the year when the 20th National Congress of the Communist Party of China will be held. At an important moment when China has entered the significant stage of building a modern socialist country and marching towards the second centennial journey, the Chinese Astronomical Society has celebrated its centennial birthday.

In the past decade, the national member congress of the Chinese Astronomical Society was held in 2014 and 2018. The Standing Council or the Council has been held for more than 50 times to jointly study and formulate the plan of the society. Through the unremitting efforts of the 12th, 13th, and 14th councils, the self-construction of the Chinese Astronomical Society has made significant progress. Ten unit members have been added, including the Department of Astronomy of Xiamen University, the Department of Astronomy of Shanghai Jiaotong University, the School of Physics and Astronomy of Sun Yat-Sen University, the Institute of Optics and Electronics of Chinese Academy of Sciences, Suzhou Teenager Astronomical Station, the School of Physics of Huazhong University of Science and Technology, Shanghai Science and Technology Museum, Yunnan University, Xihua Normal University, School of Space Science and Physics of Shandong University. More than 50 individual members add into the Chinese Astronomy Society every year, and it has established business guidance relations with Shandong Astronomical Society and Chongqing Astronomical Society. At present, the Chinese Astronomical Society has 33 unit members and more than 3,300 individual members, and has established business guidance relationships with 17 provincial astronomical societies and 2 provincial astronomical groups. On the basis of the original working committees of organization, education, popularization, term approval, and book information publishing, the youth working committee, the female astronomer committee, and the information working committee were successively established from 2016 to 2019. So far, the Chinese Astronomical Society is composed of 8 working committees and 11 professional committees, including astrometry, time and frequency, sun and heliosphere, planets, stars, galaxies and cosmology, astronomical instruments and technology, radio astronomy, space astronomy and high-energy astrophysics, astromechanics and satellite dynamics, and the history of astronomy, as well as the fund working committee.

Through the joint efforts of the members of the professional and working committees of the astronomical society, in the past decade, all branches of astronomy in China have made remarkable achievements, and made important progress in the development of a series of major astronomical equipment and the research of scientific problems. It has obvious advantages in theoretical model and computer simulation of large-scale structure of the universe, large-scale spectrum of the Milky Way, solar structure and explosion mechanism, deep space exploration and celestial bodies in the solar system. Following the LAMOST telescope, and the completion of other new major scientific infrastructure projects, such as FAST, LHAASO, and the launch of the space astronomy instruments such as Monkey King (Dark Matter Particle Explorer), Insight (Hard X-ray Modulation Telescope), and CHASE (Chinese Hα Solar Explorer), the output of major scientific research achievements has been effectively promoted. A total of 11 research papers have been published in Nature and Science journals. Such a solid foundation is made for China to further become a powerful astronomical country. Over the past decade, it has won three national second prizes in natural sciences, including the discovery of high-energy electron cosmic ray spectrum exceedance, the evolution of large samples of stars and the formation of special stars, and the study of the spiral arm structure of the galaxy based on high-precision maser astrometry, as well as 54 other national, provincial, and ministerial collective and individual awards. For example, the selection of astronomers as Models of the Times and the "May 4th" collective award shows that the influence of astronomers in China has increased significantly; and female astronomers have won National Advanced Workers and three Chinese Young Female Scientists' awards, indicating that the status of female astronomers has increased significantly.

In the past decade, China's astrometry has made many important achievements in basic research and application. The influence of the long-term optical aberration caused by the motion of the solar system around the Milky Way on the extragalactic reference frame is systematically analyzed and studied. The value of the long-term optical aberration is accurately calculated from the VLBI observation data of more than 40 years, which makes a significant contribution to the establishment of the third generation international celestial reference frame (ICRF3). At the same time, we have independently compiled an absolute proper motion catalogue in the outer sky area of the silver disk, which covers the sky area of 22000 square degrees outside the silver disk (beyond plus or minus 30 degrees of silver latitude). The position and absolute proper motion of more than 100 million celestial bodies with brightness of 20.8 mag. in the dark to R band are calculated, and the photometric data in seven bands from optics to near infrared are included. Jointly developed a high-precision negative scanner and completed the digitization of 30,000 astronomical historical negative films with a time span of nearly 100 years. The relevant results have been incorporated into the China Virtual Observatory.

The time and frequency research has a vigorous development in the past decade. With the deployment and implementation of national major projects such as Beidou satellite navigation system, national major scientific and technological infrastructure - high-precision ground-based time service system, and space station high-precision time and frequency experimental system, China has developed rapidly in the field of time and frequency, significantly enhanced its ability of independent innovation, and achieved a series of gratifying results. The world time no longer depends on foreign data. In 2019, an earth rotation measurement and data service system integrating digital zenith tube, VLBI2010, iGMAS, and other systems were built in National Time Service Center, and they provided world time data for major space programs such as Chang'e-5 and Tianwen-1. Relying on the Luonan 40m telescope, China has built a pulsar timing observation system, established a comprehensive pulsar time scale using the relevant observation data of 15 millisecond pulsars in seven years, and carried out experimental research on pulsar time controlling atomic clock. After decades of tackling key problems, the research and development of domestic small cesium clocks have finally made a series of breakthroughs, successfully developed atomic optical pumping and optical detection and optical pumping small cesium clocks, and formed a batch production capacity. The equipment is applied to Beidou navigation, navy navigation, 5G communication, electric power operation, rail transportation, and Arctic scientific research, ensuring the independent control of the operation of important infrastructure, such as the generation and transmission of national standard time. At the same time, China has independently developed 8 cubic centimeter small chip atomic clock products with the same main performance as American products and only half its volume, and successfully equipped and applied in micro autonomous navigation positioning and timing terminal, Beidou navigation special receiving terminal, underwater navigation system, and one communication system. In addition, the self-developed passive hydrogen atomic clock was also successfully operated by the Beidou navigation satellite, and the long and short wave timing system appeared in the achievements exhibition of the 70th anniversary of the founding of new China, and the Party History Museum of the Communist Party of China.

In the past decade, China has become an important research force in the international solar physics group. In terms of scientific research, China publishes about 250 articles in international journals every year, second only to the United States. According to the bibliometric statistical report of astronomy in 2014-2018 issued by the Chinese Academy of Sciences, among the 1% and 10% highly cited papers in the world the proportion of solar physics papers from China in the world is 2-3 times higher than the average level of Chinese astronomy. Influence index (the ratio of the citation frequency of articles of a specific discipline in a country to the citation frequency of articles of the discipline worldwide is 0.997, indicating that the citation of Chinese solar physics papers has reached the international average level. Through the analysis of domestic appearance measurement data or simulation, Chinese scholars have studied the basic radiation and dynamic process of the solar atmosphere, the formation of the fine structure of the solar atmosphere, the transmission of related waves and energy, and the physics of energy storage and release of solar explosion activities. A lot of progress has been made in mechanism, long-term evolution of solar magnetic field and simulation of generator. These advances have increased human understanding of basic physical processes such as radiation magnetohydrodynamics (or magnetized plasma) occurring on the sun, and provided new understanding for accurate prediction of disastrous space weather. In terms of equipment, CHASE solar satellite has been launched and the advanced space-based Solar Observatory will be launched soon, which will further enhance China's international status.

Chinese astronomers have made important achievements in planetary geodynamics, the observation and research of small celestial bodies in the solar system, the research of extrasolar planetary systems, and astrochemistry. They have been selected as the top ten astronomical scientific and technological progresses in China for five times. It is found for the first time that there is a significant amplitude enhancement signal with a period of about 8.6 years in the diurnal variation of the earth's rotation parameters, which reveals that there is a close correspondence between the extreme moment of the oscillation and the rapid change of the geomagnetic field; 27 new near earth asteroids have been discovered, including 5 asteroids that pose a potential threat to the earth. A small object light variation database has been established to reveal the thermophysical characteristics of the main belt population. It has accurately determined that the orbit of Tutattis asteroid, so that Chang'e-2 can realize kilometer flyover detection. Based on Chang'e-2 detection data, reveal the physical characteristics, rotation characteristics, internal structure and formation mechanism, surface geological characteristics and evolution history of Tutattis. Based on the LAMOST spectral data and the characterization of exoplanets and their host stars, a new hot Neptune population is found. Hundreds of extrasolar candidates were found by using the CSTAR and AST-3 telescopes in Antarctica. Using Chang'e-5 sample to carry out mineral petrological research, a relatively rare lunar high titanium basalt is found, which provides accurate evidence for in-depth understanding of the late lunar magmatism and the material geochemical characteristics of the internal lunar mantle source area.

With the completion and successful operation of a number of major national scientific research equipment, such as LAMOST, FAST, HXMT, China's research in the field of stars and the Milky Way has achieved rapid development, realizing the full band coverage of observation means from high energy to radio, and the overall research level is close to the international advanced level. During this period, Chinese researchers in the field of stellar physics published more than 2,000 articles. Using LAMOST to survey the sky, the world's largest spectral library of handed down stars was obtained. Based on the observation of the eye of insight, a large number of observation data of neutron stars, black holes, and gamma ray bursts were obtained. The fundamental frequency cyclotron absorption line of the neutron star with the highest energy, the relativistic jet closest to the black hole, and the first X-ray counterpart of the fast radio burst were found, and it was confirmed that it came from a Magnetic spin star in the Milky Way. A large number of pulsars have been discovered by FAST, which makes China's research on fast radio bursts and neutral gases in the forefront of the world. Four high dispersion new fast radio bursts were found, the polarization of repeated bursts was time-varying, the flow of the first Galactic storm source was strictly limited, and the largest fast radio burst event sample so far was found.

Chinese astronomers have made important achievements in cosmological theoretical models, dark matter detection experiments, and cosmological observation limitations. In terms of theory, we have carried out theoretical research on the origin model of the universe (especially the rebound model and inflation model), a variety of dark energy models, modified gravity models, the generation mechanism of primary black holes and gravitational waves, the early phase transition of the universe, neutrino cosmology, and so on. In terms of experiments, Jinpingshan underground laboratory was built, and two dark matter direct detection experiments, panda-x based on liquid xenon and CDEX based on high-purity germanium, were launched to obtain the parameter limits of international advanced dark matter model. Chinese astronomers have completed a series of numerical simulations with international leading status, among which pure dark matter simulation includes "Pangu", "Jiutian", ELUCID, Cosmic Growth, Cosmic Zoom, "Phoenix", and TianNu. Based on the current X-ray observations, the existence of hot gas on a large scale in the Milky Way was pointed out for the first time. Based on the data of DAMPE satellite, accurately measure the nuclear energy spectra of cosmic ray positrons, protons, and helium, discover the new structure of the energy spectrum, and obtain the most sensitive search results for gamma ray line spectrum radiation. The first batch of ultra-high-energy photon accelerator and gamma ray accelerator of Super-high-energy PEV has been found by LHAASO, to open the era of Super-high-energy astronomy.

In the field of astronomical technology and

method research, China has independently and successfully developed a number of telescopes and scientific terminals for different observation objects, and covering different observation bands, such as the terminals of LAMOST and FAST, the sky survey module of the China Space Station Engineering Sky Survey Telescope (CSST, aperture 2m), etc. Chinese Large Solar Telescope CLST with 1.8-m aperture had achieved the first light and in debugging. It has developed international cutting-edge THz antenna and detection terminal technology, including 1.4 THz band highest sensitivity super thermal electron mixer, and China's first THz superconducting camera. Xinglong 2.16m telescope and Lijiang 2.4m telescope are equipped with high-resolution optical fiber spectrometer. The development and use of these telescopes and scientific instruments not only laid the foundation for the long-term development of astronomy in China, but also improved the innovation ability of independently developing large telescopes and terminal equipment in China.

The 500 meter aperture spherical radio telescope (FAST) was completed in September 2016, passed the national acceptance in January 2020, and officially opened to the domestic market. Under the requirements of "early achievements, more achievements, good achievements, and great achievements", a number of major scientific research achievements have been made: through the original neutral hydrogen narrow line self-absorption method, FAST has made great progress in measuring interstellar magnetic field; FAST has made a series of important discoveries in the emerging research direction of fast radio bursts, including 4 cases of high dispersion new fast radio bursts, polarization time-varying of repeated bursts, strictly limiting the flow of the first Galactic storm source, the largest sample of fast radio burst events so far; FAST has found about 500 pulsars in total, etc. The newly built 65m radio telescope has greatly enhanced the ability of CVN to accurately measure the orbit of deep space probes in real time, and has successfully completed the relevant tasks of the lunar exploration projects "Chang'e-2", "Chang'e-3", "Chang'e-4", "Chang'e-5" and the Mars probe "Tianwen-1". At the application level of national demand, China VLBI Network (CVN), composed of 25 meters in Sheshan, 26 meters in Urumqi, 50 meters in Miyun, and 40 meters in Kunming, has played an irreplaceable role in China's deep space exploration projects.

In the past decade, China has made great progress in astronomy education, astronomy history, astronomy journals, astronomy validation, informatization, young astronomers, and female astronomers. The decade from 2012 to 2021 is the most rapid development of astronomy education in China. At present, 34 colleges and universities have carried out scientific research related to astronomy, and 22 Astronomy (Science) departments have been established to carry out undergraduate astronomy education, of which 12 have been established in the last decade. Astronomical science popularization participated in national science and technology activity week, national science popularization day, and other activities. In 2020, the "looking up at the stars" science night and the major demonstration activity of the science and technology activity week - "the Belt and Road" science popularization exchange week were held, and in May 2021, the academic exchange activity "science popularization of space astronomical exploration of the Chinese Astronomical Society" was held.

Looking forward to the future, the Chinese Astronomical Society has gone through a hundred years, and is moving forward to the next hundred years. Combined with the national development strategy, China will enter the ranks of moderately developed countries by 2035. Chinese astronomy will not and cannot fall behind. With the continuous progress of science and technology, and the continuous innovation and efforts of Chinese astronomers, more new technologies, methods, and equipment will continue to be invested in astronomical research, such as, the upcoming launch of ASO-S, the ground equipment includes the national major scientific research instrument development project "2.5m Wide-field and High-resolution Solar Telescope" (WeHosT), thc 2.5m Wide Field Survey Telescope (WFST), the 2m Ring Solar Telescope, the Muztag Ata 1.9m telescope, and the 1.6m Multi-channel Photometric Survey Telescope (Mephisto) in Lijiang. In addition, major projects such as China's large optical infrared telescope, 6.5m Multiplexed Survey Telescope (MUST), LAMOST phase II project, China Giant Solar Telescope (CGST), the SiTian Project, 110m fully steerable radio telescope, and 15m submillimeter wave telescope are being actively promoted, and the excellent sites of Antarctic Dome A and Western China have been confirmed. We will promote the development of China's astronomy and realize that China's astronomy will enter the international developed level as soon as possible.

Astrometry

Tang Zhenghong

In the past 10 years, many important achievements in astrometry field have been achieved in both research and application.

In terms of the astronomical reference system, the influence of the long-term aberration effect caused by the movement of the solar system around the Milky Way on the extragalactic reference frame was systematically analyzed and studied, and the long-term aberration value was accurately calculated from the VLBI observation data of more than 40 years. Significant contributions have been made to the establishment of the third generation of the International Celestial Reference Frame (ICRF3). Using the new-generation geodetic VLBI observation data, based on the phase closure constraints, the structures and their changes of some extragalactic radio sources are solved. With the help of the latest Gaia catalogue, the characteristics of ICRF3 are comprehensively analyzed, and independent verification and supplementary analysis are provided for the relevant parameters of ICRF3. Statistical analysis of the deviation of the optical and radio positions of the extragalactic source gives direct evidence that this deviation is related to the radio structure of the extragalactic source (especially, the jet structure), providing a first-hand connection to the optical and radio reference frames. The misunderstandings due to the ellipse term of aberration in the process of galactic coordinate system conversion and application are clarified, and the problems caused by the reference system conversion defined by IAU 1958 are pointed out. For the first time, it is proposed to use the modern large-scale sky survey catalogs to directly construct the coordinate system most consistent with the structure of the Milky Way, and the method of establishing the orthogonal coordinate system is discussed based on the observation data of the galactic plane and the galactic center. The IAU1958 coordinate system will produce a pointing difference of about 0.5°; this result may promote the formation of new IAU resolutions proposed by Chinese scholars in the future. Under the framework of general relativity, the concept of the ecliptic is abandoned, and the latest solar system ephemeris is adopted to obtain a set of high-precision long-term precession parameters.

In the aspect of the earth's rotation, the earth shape factor J2 calculated based on the satellite's laser ranging data was innovatively applied to the calculation of the earth precession model, and the deceleration effect of the earth shape factor J2 was considered when integrating the precession equation. As well as the self-consistency of the celestial and terrestrial reference frames can provide important references when the IAU discusses new standard precession models in the future. In terms of high-precision astrometric data processing, inspired by the high-precision astrometric calibration of the Hubble Space Telescope, a self-calibration method is proposed to accurately solve the geometric distortion of CCD images. It has been applied to the high-precision calibration of several major domestic large telescopes, and has significantly improved the astrometric accuracy (including natural satellites, near-Earth asteroids, and other celestial bodies). In addition, it is proposed that an astrometric correction method for accurately solving the poor color refraction by using the color index of the Gaia star catalog, and an accurate solution method for the distance between multiple CCD chips. These methods can be applied not only to the high-precision astrometry of ground-based telescopes, but also to the high-precision astrometry calibration of the Chinese Space Station Survey Telescope (CSST) to be launched in near future. In the astrometric measurement of ground-based telescopes, it has also successfully revealed the characteristic range with improved measurement accuracy of the positions of two close celestial bodies, which will play a guiding role in the observation of small celestial bodies in the solar system. In addition, a method to accurately measure the geocentric distance of the target with the sparse observation data of two adjacent nights of asteroids is also proposed. The new method will have good application value for the observation of newly discovered asteroids and near-Earth asteroids.

The system carried out CCD observation of natural satellites of giant planets (Jupiter, Saturn, Uranus, and Neptune), obtained a large amount of CCD image data, and solved the high-precision astrometric positions of a number of natural satellites, which were used by JPL in the United States to compile DE440/DE441 ephemeris. The orbits of natural satellites such as Triton were improved by using the orbit calculation program developed by ourselves. In cooperation with British and French scholars, the reduction and data processing of the Cassini detector observation images have been carried out, and good progress has been made.

Using the original principle of relative simultaneity observation, based on Gaia's measured data, to calculate the change of basic angle of two directions, cross-comparison with other monitoring results, so as to verify and obtain the basic angle change independently, and provide key reference to Gaia's final scientific data correction.

An absolute proper motion catalogue of the sky area outside the galactic disk has been independently compiled. The catalogue covers the sky area of 22,000 square degrees outside the galactic disk (plus or minus 30 degrees of galactic latitude), and the catalogue gives positions and absolute motions of more than 100 million celestial objects down to magnitude 20.8 at R-band, and includes photometric data in 7 bands from optical to near-infrared. This work has been awarded the top ten progresses in Chinese astronomy. The astrometric support system of the Lunar Ultraviolet Telescope (LUT) (including the guide star tracking and observation strategy) of the Chang'e-3 mission in Chinese lunar exploration project is carried out successfully. In December 2012, after Chang'e-3 landing on the moon, the Lunar Ultraviolet Telescope successfully realized the first fully automatic telescope that can be debugged and on duty unattended based on this set of software, and obtained a large amount of very valuable observation data.

With the support of the Ministry of Science and Technology's basic special project of science and education, a high-precision plate digitizer was jointly developed to complete the digitization of 30,000 astronomical historical plates of China covering about 100 years. The relevant results have been included in the China Virtual Observatory. With the support of the Shanghai Municipal Science and Technology Commission, two airborne rail digitizers have been developed independently, one of which has the function of color and reflection scanning. With the support of the Ministry of Science and Technology, nearly 20,000 astronomical historical plates from the Turin Astronomical Observatory in Italy and the Ulugh Beg Astronomical Institute in Uzbekistan have been digitized.

Time-Frequency Study and Application

Zhang Shougang, Li Xiaohui

The research of time and frequency has developed vigorously for the past ten years. With the deployment and implementation of major national projects such as Beidou navigation satellite system, high-accuracy ground-based time service system of major national scientific and technological infrastructure, and high-precision time and frequency experimental system in Chinese space station, the realm of time and frequency in our country has developed rapidly with significantly enhanced independent innovation capability, achieving a series of gratifying results.

Universal Time no longer relies on foreign data. Universal Time describes the earth's spatial position which is indispensable in space science and national defense, etc. Since the 1990s, after the traditional Universal Time measurement methods were eliminated, Universal Time in our country had completely relied on the foreign data, which posed security risks. National Time Service Center of CAS, Shanghai Astronomical Observatory, National Bureau of Surveying and Mapping, and other institutes have carried out relevant researches. National Time Service Center has completed the construction of the earth rotation measurement and data service system in 2019, which integrates digital zenith tube, VLBI2010, iGMAS, etc., and provides Universal Time data for major space programs, including Chang'e-5 and Tianwen-1.

A breakthrough has made in pulsar time. Radio telescopes, including the 25 meters in Xinjiang, 40 meters in Kunming, 65 meters in Shanghai, and 40 meters in Luonan, all have acquired the pulsar timing observation capability. The 500-meter radio telescope in Guizhou has discovered nearly 300 pulsars, and is conducting millisecond pulsar timing observation research. National Time Service Center completed the construction of the pulsar timing observation system based on 40-meter telescope in Luonan, established assemble pulsar time scale by using 7-year observation data of 15 millisecond pulsars, and conducted experimental research on steering atomic clock with pulsar time. Institute of High Energy Physics of Chinese Academy of Sciences and China Aerospace Science and Technology Corporation carried out space X-ray pulsar navigation observation experimental research.

Domestically manufactured small cesium clock has broken the monopoly of the developed countries. Due to the complex scientific and technological difficulties, time-keeping cesium atomic clock had been monopolized worldwide by US magnetic state-selection cesium beam atomic clock for half a century before 2017. US have completely banned the sales to our country since 2018. After decades of research, domestically manufactured cesium atomic clock has finally achieved a series of breakthroughs. Lanzhou Institute of Physics has realized the magnetic state-selection cesium beam clock. Peking University together with Hubei Hanguang Co., Ltd. etc. has realized magnetic state-selection optically detected cesium beam atomic clock. National Time Service Center has cooperated with Chendu Tianao Electronics Co. and China Electronics Technology Group on the development of the optically pumped and optically detected cesium beam atomic clock, and has formed mass production capacity. The performance of the optically pumped cesium beam atomic clock exceeds that of the US product. This equipment is applied to Beidou navigation, naval navigation, 5G communication, power operation, transportation, and arctic scientific research system, etc., ensuring the independent and controllable operation of the important infrastructure, such as the national standard time's generation and broadcasting, which is recommended by BIPM to generate the standard time for other countries. These products have contributed to the generation of UTC and weight in the computation of TAI.

Self-developed world's smallest atomic clock has broken the foreign monopoly. The micro atomic clock with the same size, weight, and power consumption as chips is a high-performance time and frequency signal-generating device, based on the integration of quantum physics theory and micro-electro mechanical system (MEMS) processing technology. As a component, it provides standard time and frequency signal reference for the operation of related equipment. For a dozen years, US have exclusively monopolized the international market. In recent years, US have begun to restrict the export of the chip-scale atomic clock to China. National Time Service Center of CAS and Innovation Academy for Precision

Measurement Science and Technology have organized the forces to independently develop an 8-cubic-centimeter chip-scale atomic clock whose main performance is the same as that of the US product, and whose size is only half of that of the US product. It is successfully installed in the micro-autonomous PNT terminal, Beidou navigation special receiving terminal, underwater navigation system, and a certain communication system, etc. The team has also realized a prototype with the same main performance and a volume of 2.3 cubic centimeters. This research result has been selected into the national "13th Five-year Plan" scientific and technologic innovation achievement exhibition.

The research of the optical atomic clock has entered the leading position internationally. Based on its excellent performance, the optical atomic clock will replace the cesium-based microwave frequency clock as the frequency standard to redefine the time unit "second", which is prospective. National Institute of Metrology, East China Normal University, Huazhong University of Science and Technology, Innovation Academy for Precision Measurement Science and Technology of CAS, and National Time Service Center have successively developed optical atomic clock with different elements, all of which have made different progresses and have entered international leading positions.

The world's first cold atom microwave clock is operated in space. Based on the principle of the ground-based cold atom fountain clock and microgravity in space station, the cold atom microwave clock in uniform rectilinear motion developed by Shanghai Institute of Optics and Fine Mechanics of CAS has successfully operated on Tiangong-2 space lab. High-precision time and frequency experimental system in Chinese space station, including the space station time and frequency generation system, space-ground time and frequency comparison system, and ground testing and application system, has entered the stage of developing the formal prototype. In the near future, the world's first cold atom optical clock will be operated in space. The passive hydrogen atomic clock developed by Shanghai Astronomical Observatory and China Aerospace Science and Industry Corporation has successfully operated on Beidou navigation satellite. The Beidou navigation satellite system with the advanced performance has begun to provide navigation, positioning, and timing services to the world.

The long-wave and short-wave time signal

systems were exhibited in the 70th Anniversary Achievement Exhibition of the Founding of the New China and The Chinese Communist Party History Museum. With the development and application of microwave frequency standard, time-keeping clocks, and methods like joint timekeeping in different places, improvement of timekeeping algorithms, and optimization of international time comparison, the divergence of UTC (NTSC) with respect to UTC has been improved from 10ns to less than 5ns, and the weight contributed to international standard time has been improved from the fifth rank to the third rank in the world, which provides a reliable and stable time reference for the nation's long-wave, short-wave, Beidou and Changhe time service systems. The multi-method time service systems including telephone, network, optical fiber, short-wave radio, long-wave radio, and satellite have already been established, which has better supported economic and social operation and national security development.

Time and frequency measurement comparison instrument has achieved technological breakthrough. The dynamic range and precision of time interval measurement instrument and the self-noise of the frequency stability analyzer have led the world, but as a general measurement instrument, its versatility and environmental adaptability need to be further improved. National Time Service Center is carrying out the research and the production through cooperation with enterprises; the precise optical frequency measurement is a hot spot and trend of the application and development of time and frequency measurement. The main performance of the optical frequency synthesis and measurement equipment based on femtosecond frequency comb developed by National Time Service Center is internationally advanced, and can be applied to different high-precision measurement fields. The application of self-developed ultra-stable laser system and femtosecond frequency comb has realized the ultra-stable optical-cavity-based microwave source, and the relative frequency stability has been improved by two orders of magnitude than that of the normal ultra-stable microwave source. Its application improves the short-term frequency stability of the cold atom fountain clock in the National Time Service Center by half an order of magnitude. Instruments such as satellite common view, two-way satellite based time comparison, long-wave timing, and remote time comparison and reproduction have been made domestically, and have been applied to major facilities. The Eurasian time comparison technology based on Beidou satellites has been recommended by international organizations for international time comparison.

The technology of high-precision time and frequency transfer has developed rapidly. Optical fiber is the highest-precision method currently available for time and frequency transfer among distant sites. By using the communication carriers' optical fiber, thousand-kilometer high-precision time and frequency transfer can be realized, including optical frequency signal, microwave frequency signal, and time signal. Quantum clock synchronization has the characteristics of high accuracy and strong security. National Time Service Center has achieved sub-picosecond time synchronization stability by using 50-kilometer optical fiber, laying the foundation for long-haul quantum clock synchronization application. Based on optical fiber time and frequency transfer technology and enhanced Roland time signaling technology, National Time Service Center has undertaken the research and construction of the "high-accuracy ground-based time service system" of major national scientific and technological infrastructure.

The continuous development of the society requires more for time and frequency technology and service. In the future, with the construction and operation of major facilities, such as space-based time and frequency system, high-accuracy ground-based time service system, and so on, China will achieve a higher-performance, more resilient and more integrated national time standard generation and service system, which will strongly support the construction of a powerful country in science and technology, and high-quality economic and social development.

Solar Physics

Ji Haisheng

In the past ten years, solar physics research team has grown continuously in China, which is now playing an important role in the international solar physics community. In terms of research papers, the country publishes about 250 articles in international journals every year, second only to the United States. The 2014-2018 bibliometric statistics report in the field of astronomy issued by the Chinese Academy of Sciences shows that in the global 1% and 10% highly cited papers the proportion of solar physics papers from China in the world is 2-3 times higher than the average level of Chinese astronomy. The influence index (the ratio of the average citation frequency of a specific subject in a country to the global average citation frequency of the subject) is 0.997, indicating that the citations of Chinese solar physics papers have reached the international average level. Through the analysis of domestic and international observational data or simulations, Chinese scholars have made a lot of progresses in the field of the basic radiation and dynamic processes of the solar atmosphere, the formation of the fine structure of the solar atmosphere and the transmission of related waves and energy, the physical mechanism of energy storage and release in solar eruptions, the long-term evolution of the solar magnetic field, and the simulation of dynamo process. These advances have increased human understanding of basic physical processes such as radiative magnetohydrodynamics (or magnetized plasma) that occur on the sun, and provided new insights for accurate disastrous space weather forecasting. It is worth mentioning that most of the work is led by young scholars.

In ground-based observations, Huairou solar observatory and Fuxian lake solar observatory are under normal operation, the full-disk telescopes of magnetic field and Hα given by the meridian project will be installed at Ganyu solar observing station soon. From 2014, we have seen more than 100 referred papers being published using the observations made by the New Vacuum Telescope of Fuxian lake solar observatory. The representative progress includes the direction observations of magnetic reconnection the Chromosphere, fast plasmoid-mediated reconnection in a solar flare, twisted magnetic structure caused by the rotation of a pore or network magnetic fields and its triggering role in solar activities, and high-resolution images of solar prominences etc. A breakthrough in corona observations has been made. China successfully developed several sets of millisecond-level high-resolution dynamic broadband radio spectrometers, which are installed at the National Astronomical Observatory, Yunnan Astronomical Observatory, and Shandong University. Based on the data accumulated by broadband dynamic spectrometer in the past 20 years, a number of unique microwave spectral fine structures, spike groups and flare radio precursors have been discovered, and a new method for diagnosing the magnetic field and particle acceleration process in the source region of the explosion has been developed. MUSER (MingantU SpEctral Radioheliograph) is a solar-dedicated interferometric array that is located in Mingantu town, Inner Mongolia of China and carries out imaging spectroscopy of the Sun with high spatial resolution, high time resolution, and high frequency resolution at a wide frequency range. Under the framework of Meridian project phase II, meter-wave and ten-meter-wave radioheliographs are under construction at Mingantu and Daocheng, covering the spectral range of 30–450 MHz. In the past ten years, we have carried out the site survey for future ground-based observation in Western China, and selected the site of Wumingshan in western Sichuan for the proposed large-aperture ground-based solar telescope (including the coronagraph). Several referred papers have been published on the results of the site survey. Qinghai Lenghu was selected as the site for infrared solar observation. We have finished scientific aims for the 8-meter solar telescope, and we are still applying for official recognition and financial support. The ground-based 2.5-meter large field of view solar telescope and 2-meter ring solar telescope are under construction.

In space-based observations, Hα spectral telescope (CHASE) has been launched successfully and has obtained scientific data, this marks a substantial breakthrough in the space observation of our country's heliophysics. Meanwhile, the Advanced Space-Based Solar Observatory (ASO-S) will be launched in 2022, and its scientific application system has been basically built. Several space (pre-research) projects have been proposed, mainly focusing on stereoscopic observations deviating from the Sun-

Earth line, including stereoscopic or panoramic observations on the ecliptic plane, and flying away from the ecliptic plane in the direction of the solar poles. In addition, innovative projects proposed by Chinese scholars, such as ultraviolet spectrometer observation of solar (star) coronal magnetic field and stereoscopic magnetic field observations, have also been proposed and fully pre-researched.

In space weather prediction, many domestic units have formed corresponding research teams, and the empirical prediction model and numerical simulation prediction model have achieved certain results. In recent years, the amount of the observational data has increased rapidly, and more and more intelligent technologies have been used to extract precursory features and achieve good prediction results. Scholars have successfully used the "machine learning" method to predict the coronal Soft-X-Ray radiation from ultraviolet radiation data. In CME's earth effectiveness prediction, a great progress has been made in recent years. Recent research shows that the prediction of CME landing time by machine learning method shows good results, and the error is reduced to about 6 hours, which is similar to artificial prediction. The research also shows that the stereocopic observation of solar bursts will further improve the prediction accuracy.

In the study of stellar activity, through years of efforts, Chinese solar physicists have kept pace with the international mainstream. Representative works include: (1) based on the stellar light variation data of Kepler and Tess space telescopes, the activity indexes of stellar magnetic field and flare are established, compared with solar activity, the stellar flare activity data set is constructed, and the statistical distribution law of stellar flare activity is analyzed; (2) Based on the stellar spectrum observation data of LAMOST telescope, the stellar chromosphere activity spectrum data set is constructed; (3) the statistical law of stellar chromosphere activity is obtained, and the relationship between stellar chromosphere activity and stellar age is analyzed. The "main" sky survey observation equipment represented by LAMOST telescope and its massive observation data are the advantages of China in this research field.

This report mainly refers to the *Solar Physics Planning* (2020) written by Wang Jingxiu, Chen Pengfei, Tian Hui, etc.

Planetary Science

Ji Jianghui, et al.

In recent ten years, Chinese astronomers have made significant achievements in planetary geodynamics, small bodies in the solar system, exoplanets, and astrochemistry etc. Roughly 300 papers are published in high-level academic journals (*ApJ* series, *A&A*, and *MNRAS*) , including 1 paper in *Nature*, 3 papers in *Nature Astronomy*, 1 paper in *Nature Communications*, and 3 papers in *PNAS*. The related references have been selected as one of top ten achievements of astronomical science and technology (TOP10AST) in China for 5 times.

Planetary geodynamics: It is innovatively discovered that there is a significant amplitude enhancement signal with a period of about 8.6 years in the change of the day length of the Earth's rotation, indicating that the extreme moment of the oscillation has a close correspondence with the occurrence of rapid changes in the geomagnetic field. The self-consistent internal structure model of Jupiter and Saturn as the deep structure of atmospheric circulation is established by integrating the high-precision gravity field measurement between JUNO and Cassini. The mechanism of Jupiter's diluted core structure is proposed to its internal and formation. It is revealed that the burial depth of the strongly magnetized region of the northern hemisphere lithosphere is shallower, while that of the strongly magnetized region of the southern hemisphere lithosphere is much deeper through the local spectral analysis method. It is systematically analyzed the hydrodynamic phenomena of liquid cores driven by both precession and thermal instability of the terrestrial planet's rotation axis, and discovered the localized thermal convection conditions.

Small bodies in the solar system: A series of achievements have been made in near-Earth objects detection, physical properties, and monitoring and warning etc. There are 27 new near-Earth asteroids discovered, 5 of which are potentially dangerous to the Earth. A light-curve database of the small bodies is established based on the observations. The thermophysical properties of the main-belt population are revealed. Precise determination of Toutatis' orbit had made Chang'e-2 achieve a kilometer-level flyby. The physical properties, rotation properties, internal structure and formation mechanism, surface geological features and evolution history of Toutatis are further revealed based on the images shot by Chang'e-2's flyby, which are selected as TOP10AST in China in 2012 and 2013, respectively. The 3D shape evolution model is employed to analyze the formation of the flattened shape of Arrokoth, and proposed to unveil its shape formation and evolution mechanism of small icy bodies under the effect of solar illumination induced by mass loss, thereby revealing the formation history of planetesimals in the Kuiper Belt. The morphological features of the mid-infrared moon and its changing laws with the phases of the moon are obtained to reflect its physical nature. The lunar eclipse is investigated with Gao-fen satellite to provide the characteristics of lunar eclipse luminosity changes and multiple hot spots. In situ lunar space weathering gives clues to the cause of the decrease in spectral slope caused by weathering and thermal radiation with wavelengths less than 2.3 μm, implying the characteristics of micro-scale lunar thermal radiation.

Exoplanets: Using the data of LAMOST and Kepler, the eccentricity distribution feature of exoplanets is revealed, and the planet formation scenario is further constrained, which was selected as TOP10AST in China in 2016. Moreover, based on LAMOST spectral data and the characterization of exoplanets and their host stars, new hot Neptune population is discovered, being selected as TOP10AST in China in 2018. Hundreds of exoplanet candidates are discovered using the CSTAR and AST-3 telescopes in Antarctica, which was awarded as TOP10AST in China in 2019. The systematic exploration of exoplanet formation, evolution, and dynamics reveals a new mechanism of near-resonance configuration based on planetary growth, planetary disks, and gas giants etc. A trough in planet size distribution around two Earth radii induced by the atmospheric escape is found. The fine structure of the HL Tau protoplanetary disk is quantitatively characterized, which won the second prize of science and technology in Jiangsu Province. The origin of dust ring structure in protoplanetary disk, evidence of vortex structure, and characteristics of spiral arm structure generated by vortex are revealed by ALMA observation and numerical simulations. Based on the transmission spectrum observation of large-sample hot Jupiter's spectral photometry, a clear-sky feature characterized caused by the pressure of alkali metal atoms was found in its atmosphere, and the suspected signal of lithium was detected for the first time, revealing that there were sodium atoms, potassium atoms, water molecules, titanium oxide molecules, and other atmospheric components. The atmosphere of classical hot Jupiters is extensively described using a high-resolution Doppler spectroscopy, observational evidences for the heating mechanism of the upper atmosphere of different types of planets are provided.

Astrochemistry: The Chang'e-5 return sample (CE5C0000YJYX065) was investigated for mineralogy and petrology, and turned to be a relatively rare type of high titanium lunar mare basalt, which provides accurate evidence for understanding the late lunar magmatic activity and the geochemical characteristics of lunar mantle source region. The largest chlorine isotope fractionation in the solar system to date in lunar meteorites was reported. The mineralogy, geochemistry, and chronology of Martian meteorites were carried out to elucidate the eruption events of basalt magma and complex impact histories on Mars. The thermal metamorphism and impact histories of meteorites from various asteroids (including Vesta) have been studied to reveal the time and mechanism of differentiation of asteroid parent bodies. High-precision potassium (K) isotope analysis of tektites and chondrites was conducted to explore the depletion and fractionation mechanism of moderately volatile element potassium in the solar system.

Chinese astronomers have actively taken part in international cooperation, such as the dynamics of exoplanet with University of California, Santa Cruz, 3D Radiation Transfer Model of Coma with researchers of Max-Planck Institute for the solar system in Germany etc. Chinese astronomers participate in the International Joint Testing of Near-Earth Objects, and the IAWN international organization under the framework of the United Nations, representing China. They participate in international projects or observations to detect exoplanets, such as through KMTNe (microlensing), GAIA (astrometry), SPIRou (radial velocity), LCOGT (transit), GTC to explore planetary atmospheres, ALMA to characterize protoplanetary disks, SDSS to describe host stars, etc. In recent ten years, a well-structured research team has been formed through talent training and being fostered, and 6 scholars have been selected into the national talent plan. Chinese astronomers are planning space missions to detect exoplanets in the habitable zone as the primary scientific goal, such as the "Closeby Habitable Exoplanet Survey", the "Earth 2.0" and the "Purple Eye" missions, etc.

Stellar Physics

Wang Xiaofeng

Over the past decade, with the completions and successful operations of several national key scientific instruments such as the Large Sky Area Multi-Object Fiber Spectroscopic Telescope (LAMOST), the Hard X-Ray Modulation Telescope (HXMT), the Five-hundred-meter Aperture Spherical radio Telescope (FAST), China's research on stellar physics and the Galaxy has achieved rapid developments, to approach the world advanced level in some aspects. During this period, researchers in the field of stellar physics published 2 papers in *Science*, 8 papers in *Nature*, nearly 20 papers in *Nature* sub-journals, and around 2000 articles in professional first-class journals (including the *Astrophysical Journal* series, *Astronomy & Astrophysics*, and *Monthly Notices of the Royal Astronomical Society*). Among the 16 highly cited articles (with >200 citations) published by domestic scholars of astronomy during the period from 2011-2020, the papers in the area of stellar physics account for half of them.

Some important results include those from the LAMOST survey which has provided the world's largest library of stellar spectra. Based on this spectral database, large sample of stellar ages and the metallicity distribution were obtained; numerous stars with special abundance have been found, with their origins being better constrained. Moreover, three-dimensional picture of stellar mass distribution for the Milky Way disk and star formation history in different positions of the disk have been precisely described; and the radial and vertical metallicity gradients and distribution functions for stellar populations with different ages have been provided. The observations reveal that the Milky Way disk might have warp structure. The line of sight velocity observations led to the discovery of a massive stellar black hole candidate in the Galaxy. Combined the LAMOST spectra with the Kepler data, the red giants and the red clumps can be distinguished for the first time; white dwarfs are found to have enriched oxygen abundance in their cores; stars with super flares are found to exhibit enhanced magnetic activity. Combined with other survey data, the halo structure of the Milky Way was found to be flattened inside and nearly spherical outside; the rotation curve out to 100 kpc from the Galactic center was described, and the local dark matter density was accurately determined.

The completion and operation of the Hard X-Ray Modulation Telescope "Insight" have supplemented China's observations in high-energy physics, and significantly increased the influence of China's research in the field of compact objects. Based on the insight-HXMT data, large numbers of neutron stars, black holes, and gamma-ray bursts were obtained, resulting in numerous important results: detection of the highest-energy fundamental cyclotron resonance scattering feature in a neutron star; discovery of the nearest relativistic jet from a black hole candidate; discovery of the first X-ray counterpart of the FRB (fast radio burst) and identification of it as a Galactic magnetar; high-efficient detections of gamma-ray bursts. On the other hand, the FAST is the largest single-dish radio telescope in the world. Observations with this equipment lead to discovery of a large number of new pulsars, and make our research on fast radio burst and neutral gas stand in forefront of the world. The joint observations of the HXMT and FAST successfully constrain the origin of some fast radio bursts.

In addition to the research based on the above major instruments, Chinese researchers also have carried out unique and significant studies on molecular cloud/star formation, Milky Way magnetic field, the Milky Way spiral arms, supernova and cosmology, etc., and take a place in the world. The detailed results are as follows: (1) completing the polarization survey of the Galactic plane, probing the large-scale magnetic field in more than a half of the Galactic disk, and revealing toroidal magnetic field in the Galactic halo; (2) measuring the distance to the Perseus spiral arm of the Milky Way accurately, describing the spiral arm structure in the solar neighborhood in several kpc; (3) examining the CO completeness based on three independent CO surveys, discovering a new segment of the spiral arm; (4) obtaining extinction law of stellar and interstellar medium with the help of various sky surveys; (5) discovering hundreds of supernovae explosions in the nearby galaxies, revealing the existence of two distinct progenitor populations of type Ia supernova, improving the precision of cosmological distance measurement.

In theoretical studies, new models including

stellar convection have been developed and improved. Some long-standing problems have been solved, and the structure and evolution have been accurately determined for a number of stars. Moreover, great progress has been made in the study of stability criterion of binary interaction. Our researchers internationally pioneered and popularized the method of binary population synthesis in the study of stellar population and galaxies, achieving outstanding breakthroughs in the study of peculiar stars (such as progenitors of type Ia supernovae, hot sub dwarf, X-ray binary, etc.). Our scholars proposed that the central energy of gamma ray bust may originate from a hyper-accreting black hole or a fast rotating strongly magnetized neutron star (millisecond magnetar), which can explain the continuous activities after merger.

In the near future, with the completions and operations of a number of ground- and space-based observation facilities, such as the USTC-PMO 2.5-meter sky survey telescope (Wide Field Survey Telescope), the 1.6-meter telescope (Multi-channel Photometric Survey Telescope) from Yunnan University, China-France SVOM (Space Variable Objects Monitor), Einstein Probe, POLAR-2 from China Space Station, and Chinese Space Station Telescope (CSST), etc. China is expected to make numerous and significant breakthroughs in the main frontiers of stellar physics.

Galaxies and Cosmology

Li Cheng

In the past decade, utilizing observational facilities (both domestic and international) and numerical simulations, astronomers in China have done tons of studies in areas of galaxy and cosmology, which have been constantly promoting the international impact of the Chinese astronomy community.

Astronomers have made important progress in cosmology including theoretical models of cosmology, dark matter detection experiments, and observational constraints of cosmological parameters. Theoretical studies have covered cosmic origin models (particularly Bouncing Cosmology and inflation models), multiple dark energy models, modified gravity models, formation mechanisms of primordial black holes and gravitational waves, phase transformation in the early universe, neutrino cosmology, etc. Experimentally, the China Jinping Underground Laboratory was constructed and two experiments were initiated: Panda-X based on liquid xenon and CDEX based on high-purity germanium, both aiming for direct detection of dark matter; meanwhile the Wukong spacecraft was launched successfully, and it has measured spectral energy distributions of cosmic rays with high energy resolution and performed indirect detection of dark matter particles. Observational constraints of cosmological parameters have been improved by jointly using observational data from various experiments (e.g. Planck, BOSS, eBOSS, etc.) as well as newly developed statistics (e.g. optimal redshift weighting, multi-source cross-correlation, tomographic Alcock-Paczynski test, void statistics, hybrid modelling of redshift space distortion, deep learning, etc.). In particular, dynamical evolution of dark energy was supported at a relatively high confidence level. The FourierQuad weak lensing measuring pipeline, the first pipeline of such developed by Chinese astronomers at an international advanced level, was successfully applied to imaging surveys such as CFHTLS and DESI.

A number of numerical simulations of international advanced level have been carried out. These include dark matter-only simulations: PanGu, The Ninth Heaven, ELUCID, Cosmic Growth, Cosmic Zoom, Phenix and TianNu. With these simulations, astronomers have accurately measured the velocity bias of dark matter halos, quantitatively calculated the effect of axion dark matter on cosmic structure formation, precisely resolved the matter distribution internal to dark matter halos over 21 orders of magnitude in spatial scale, made precise prediction on the signal of dark matter annihilation on scales of galaxy clusters, and reconstructed angular momenta of both dark matter halos and galaxies from primordial density fluctuations, for the first time finding the correlation of galaxy angular momentum with the primordial density fluctuations of the universe. With international collaboration, Chinese astronomers have done the hydrodynamical simulation NIHAO, which simulates the formation and evolution of 100 galaxies with high resolutions. Taking advantage of its large sample and high resolution, NIHAO has been used to systematically study the formation process of galaxies from dwarfs to Milky Way. In addition, based on statistics of galaxy surveys, Chinese astronomers have constructed analytic models of the evolution of conditional mass functions, characterized the mass growth and star formation history of galaxies in dark halos, and obtained the most precise measurement of the halo mass for the Milky Way by jointly using numerical simulations, galaxy surveys, and the Gaia data.

Astronomers in China have accomplished several surveys for galaxies and quasars at intermediate and high redshifts. The Southern Cap U-band Sky Survey (SCUSS) and the Beijing Arizona Sky Survey (BASS) are both adopted by the international survey projects SDSS-IV/eBOSS and DESI for target selection. The Lyman Alpha Galaxies in the Epoch of Reionization (LAGER) project has detected a number of Lyman Alpha emitting galaxies at z~7, for the first time revealing the anisotropy of cosmic reionization and discovering the most distant proto-cluster of galaxies. The CFHT Large Area U-band Deep Survey (CLAUDS) carried out by an international collaboration and the narrow-band imaging surveys of both Lyman Alpha Emitters (LAE) and H Emitters (HAE) at z=2.3 carried out by Chinese astronomers alone have both provided first-hand data for detection and statistics of intermediate-to-high redshift galaxies. Furthermore, surveys based on domestic facilities such as LAMOST have identified a large number of luminous quasars and changing-look AGN at z>5, as well as the most luminous quasar in the early university with the most massive black hole discovered so far.

For galaxies at low redshifts, astronomers in China have been deeply involved in the largest integral-field spectroscopy survey, SDSS-IV/MaNGA, and have led about one third of the scientific projects which have made important progress on variation of stellar initial mass function, star formation quenching, star formation histories of both low-mass and high-mass galaxies, dust attenuation laws, dynamical models of different components of galaxies and dark matter mass fraction, counter rotation of gas and stellar components, AGN host galaxy properties, connection of star formation and AGN with galaxy-galaxy interactions, and so on. In the meantime, astronomers also have obtained multi-slit spectroscopy for nearby big galaxies using domestic telescopes, which were combined with integral-field spectroscopy data from international projects for series of studies of galaxies.

Astronomers in China have made use of domestic and international facilities to obtain multi-band observations for studies of cold gas in galaxies and circum-galactic and inter-galactic media, as well as AGN feedback. The first HI maps of several nearby galaxy clusters/groups were obtained using FAST and ASKAP, the pathfinder of SKA. A new star formation law, the extended star formation law, was proposed based on multi-band observations of cold gas of galaxies. A class of giant Lyman alpha nebulae were discovered at z=2 and for the first time astronomers observed the surface brightness of circum-galactic medium to evolve with redshift, using the integral-field spectrograph on the Keck telescope. For the first time the existence of large-scale hot gas in the Milky Way was pointed out based on current X-ray observations. High-temperature outflows were detected around the central black hole of M81, thus supporting the existence of hot accretion flow as theoretically predicted. Astronomers detected extended iron emission lines at kpc scale in AGN, as well as its association with the cold gas at galactic scales. A statistical study of the high-speed gas outflow driven by AGN was done based on a new method which makes use of QSO luminosity variation.

In addition, astronomers in China have investigated the intrinsic relationship between black holes and galaxies, as well as its cosmological implications, from a variety of viewing angles including search of intermediate mass black holes and black hole tidal disruption events, high-redshift black hole-galaxy mass relation, high-speed inflow of black hole accretion disk, outflow velocity profiles of broad absorption line QSOs, resolving the cosmic X-ray background, etc.

We see a bright future of the field of galaxy and cosmology in China, which will make another big step forward in the coming decade, thanks to a large number of current and next-generation facilities. These include both the facilities that are already in operation such as FAST and those to be in operation in the next years such as AliCPT, CSST, LAMOST-II, WFST, HUBS and MUST, as well as many international projects such as DESI, PFS, SKA, etc.

Astronomical Technology and Methods

Gong Xuefei, et al.

Over the last decade, China has made great progress in astronomical techniques and methods. A number of telescopes and scientific terminals have been developed independently and successfully, oriented to different observation objects and covering multiple wavebands. The development and application of these telescopes and scientific instruments not only laid the foundation for the long-term development of astronomy in China, but also enhanced China's innovation capabilities of independently developing large telescopes and terminal instruments.

Large Sky Area Multi-Object Fiber Spectroscopy Telescope (LAMOST), the first national large research infrastructure in China's astronomical community, started its formal spectral survey in 2012. Up to now LAMOST has obtained more than twenty millions spectra, and becomes the first spectroscopic survey project to obtain over tens of millions spectra in the world. Both of the tens of millions of spectral data released by LAMOST and the emerging breakthrough scientific research results have proved the success of LAMOST's forward-looking design concept and independent innovation technology. The Five-hundred-meter Aperture Spherical radio Telescope (FAST) has passed national evaluation and officially begun operation. FAST has become the largest and most sensitive single aperture radio telescope in the world now. Benefiting from the advantage of its ultra-sensitive, FAST has already produced and will continue to produce far-reaching scientific achievements in the fields of the pulsar searching, the origin of fast radio bursts, galaxy formation and evolution, gravitational-wave detection and so on. The Large High Altitude Air Shower Observatory (LHAASO) has entered the stage of scientific operation. LHAASO is the largest and most sensitive ultra-high-energy (UHE) gamma ray detector and cosmic ray observatory with the widest energy coverage in the world.

The Chinese Survey Space Telescope (CSST) has been officially approved, and this 2-meter space telescope will be equipped with five first-generation instruments including the sky survey module. This will be a significant opportunity for Chinese astronomical science's marching towards the international frontier. Dark Matter Particle Explorer (DAMPE), the first satellite used for space astronomical observation of China, has successfully entered orbit and started scientific observation. DAMPE has obtained the most accurate measurements of individual spectra of high energy comic ray electrons, protons, and helium above 1 TeV in the whole world so far. The Hard X-ray modulation Telescope (HXMT), also known as "Insight", is China's first space X-ray astronomical satellite, and has promoted China to occupy an important place in the field of High energy astrophysical observation. Chinese Hα Solar Explorer (CHASE), the first solar exploration science and technology test satellite, has been launched successfully. Advanced Space-based Solar Observatory (ASO-S), the first comprehensive solar exploration satellite of China, will be launched in the second half of 2022. ASO-S will realize simultaneous observation on solar magnetic field, solar flares, and coronal mass ejections in a single platform. Einstein Probe (EP), an X-ray astronomical satellite dedicated to time-domain astronomy and high-energy astrophysics, is now in the engineering development stage. China's first space passive hydrogen maser has been successfully launched, and laid a technical foundation for the improvement of the precision positioning and autonomous navigation performance for the Beidou Navigation Satellite System.

National Major Scientific Instrument and Equipment Development Project "the 2.5m Wide-field and High-resolution Solar Telescope" (WeHosT) has been officially approved. The 2.5m Wide Field Survey Telescope (WFST) in Lenghu, the 2m Ring Solar Telescope, the Muztag Ata 1.9m telescope, and the 1.6m Multi-channel Photometric Survey Telescope (Mephisto) in Lijiang are all under development. Chinese Large Solar Telescope CLST with 1.8-m aperture had achieved the first light and in debugging. The 1.2m telescope laser ranging system realized the distance measurement between the earth and the moon for the first time, which filled the gap in the field of lunar laser ranging in China. The unattended Antarctica Survey Telescope (AST3) has been put into observation in Antarctic Kunlun Station, and is the only telescope in China which has detected the optical corresponding signal of an important gravitational-wave event GW170817. The Optical and Near-infrared Solar Eruption Tracer (ONSET) and the Chinese Spectral Radio Heliograph (CSRH) have been successfully built and started scientific observation. Shanghai 65m Radio Telescope, the Asia's largest steerable radio telescope, has been put into operation. It successfully completed the VLBI orbit determination and positioning during chang'e-5 returning samples from the moon and during Tianwen-1 landing on Mars.

During the last ten years, considerable progress has also been made in the development of astronomical scientific instruments. International advanced terahertz antenna and detection terminal technology have been developed, including the highest-sensitivity superconducting hot-electron-bolometer mixer in the 1.4-THz frequency band and China's first terahertz superconducting camera. The Xinglong 2.16-meter telescope and the Lijiang 2.4-meter telescope have been equipped with high-resolution fiber spectrograph. International cooperation major project "the High-Resolution Ultra-Stable Spectrograph for GTC" will become the most competitive stellar radial-velocity measurement device in the northern hemisphere. The scientific instruments jointly developed with international major projects such as the TMT, the MSE, and the Hale telescope are also making stable progress.

Looking forward to the future, major projects such as China's Large Optical/infrared Telescope (LOT), the 6.5m Multiplexed Survey Telescope (MUST), LAMOST Phase II project, Chinese Giant Solar Telescope (CGST), the SiTian Project, the 110-meter aperture fully steerable radio telescope, the 15-meter aperture submillimeter radio telescope, and other major projects are being actively promoted. Excellent observatory sites in Dome A (Antarctica) and Western China have also been further confirmed. With the rapid development of the science and technology, through the continuous innovation and efforts of Chinese astronomical researchers, more new technologies, methods, and instruments will be invested in astronomical researches, which will effectively promote the development of astronomy in China.

Radio Astronomy

Shi Shengcai, et al.

During the last decade, radio astronomy in China has been making strides toward the forefront of the international community. Vast growth of its width, depth, and talent pool has been realized.

The Five-hundred-meter Aperture Spherical radio Telescope (FAST, a.k.a. "China Sky Eye") was built between March 2011 and September 2016. In January 2020, FAST passed the national project review and started its operation as a formal national facility. A series of major breakthroughs have been achieved. Based on an original observing technique, namely HI Narrow Self-absorption Zeeman effect (HINSA Zeeman), FAST realized the first measurement of magnetic field strength in molecular ISM through an atomic line. FAST has discovered 4 high-DM fast radio bursts (FRBs), discovered time variation of FRB's polarization, constrained the radio flux of the first known Galactic FRB, and obtained the largest set of FRB pulses ever published. With more than 500 new sources, FAST has become the leading pulsar hunter in the world. Studies of specific pulsar have also been fruitful, exemplified by the first evidence of a 3D alignment between a pulsar's spin and its spatial velocity. In addition, the successful collaboration between FAST and Fermi-LAT has resulted in systematic discovery of high energy pulsars through synergistic ground-space analyses.

Utilizing other radio telescopes in China (Tianma 65m, Ulumuqi 26m, Delingha 13.7m, etc.) and international facilities, Chinese astronomers have achieved numerous significant achievements, including mapping the sugar molecules around the Galactic center; surveying the CO, NH3, and H2CO molecules; various interstellar masers, based on high precision observations of which structures of Galactic spiral arms were revealed; the star formation on Galactic scales, in metal-poor galaxies, under the influence of magnetic field; molecular clouds on Galactic scales and their interaction with supernova remnants; dynamic and astrochemistry characteristics and Planck cores; transient search in submillimeter band; glitches, timing noise, and emission characteristics of pulsars and magnetars; Galactic electron density model YMW16; AGN light curves on long and short time scales, and etc.

The Chinese VLBI network (CVN), consisting of the Shanghai Sheshan 25m, Ulumuqi 26m, Miyun 50m, and Kunming 40m telescopes, has been an essential component of the national deep-space-exploration programs. The recent addition of the Tianma 65m has greatly enhanced CVN's accuracy in real-time-localization of deep-space probes, which helped Lunar exploration programs Chang'e-2, Chang'e-3, Chang'e-4, Chang'e-5, and Mars exploration program Tianwen-1. Technical developments, such as dual band ΔDOR and in-beam dual target VLBI, have played important roles in Lunar exploration.

In addition to FAST, a slew of construction and technical developments of radio instrumentation have been accomplished. Tianma 65m, as the first independently-developed fully-steerable Chinese radio telescope, has realized state-of-art capabilities. Qitai 110m fully steerable radio telescope (QTT) was approved at the CAS and provincial level. For the development of terahertz astronomy at Dome A in Antarctic, the conceptual design and scale-model verification of a 5-m THz telescope are accomplished, and state-of-the-art superconducting receivers and an imaging array and unattended ultra-wideband THz FTS have been developed.

As a major partner of EAO, operate JCMT. Sanctify the SKA intergovernmental treaty, built the SKA regional center pathfinder. In 2021, FAST released the first international call for proposal. Join the VLBI, EVN, and EHT observing campaigns. Establish multiple bilateral forums, including China-US Workshop on Radio Astronomy Science and Technology, Australia-China Consortium for Astrophysical Research (ACAMAR) conference series, etc.

Space Astronomy and High-energy Astrophysics

Wu Xuefeng, Wang Xiangyu, Gu Weimin, Xiong Shaolin and Fang Jun

In the past decade, with the launch of space observatories, including the Dark Matter Particle Explorer (DAMPE), the gamma-ray burst polarimeter POLAR onboard Tiangong-2 Spacelab, the Hard X-ray Modulation Telescope (Insight-HXMT), the CubeSat mission Polarimeter Light (PolarLight), the Gamma-ray Integrated Detectors (GRID), the Gravitational-wave high-energy Electromagnetic Counterpart All-sky Monitor (GECAM), and the Chinese Hα Solar Explorer (CHASE), and the construction and operation of a number of ground-based facilities, such as the Antarctica Survey Telescopes (AST3), the Five-Hundred Meter Aperture Spherical Radio Telescope (FAST), and the Large High Altitude Air Shower Observatory (LHAASO), Chinese scholars have made a series of breakthroughs in space astronomy and high-energy astrophysics.

In the field of gravitational-wave (GW) events and gamma-ray bursts (GRBs): The optical counterpart of the first GW event from the binary neutron star (NS) merger GW170817 was successfully detected by AST3 telescope in Antarctica, and Insight-HXMT provided stringent constraints on the MeV emission of the GRB 170817A; Chinese scholars participated in the observation program based on the Very Large Telescope (VLT) of the European Southern Observatory, and obtained the most adequate data of spectra and polarization for the kilonova of GW170817, and carried out comprehensive investigations on the associated GRB prompt emission, the kilonova and afterglows of GW170817; Several signals consistent with the predictions of the theoretical kilonova/merger-nova model have been discovered from the archive data of gamma-ray bursts; A special GRB transiting from the baryon-dominated-fireball to the Poynting-flux-dominated phase and the shortest ever type-II GRB were discovered; The self-organized criticality of X-ray flares in GRB X-ray afterglows was found statistically; POLAR obtained the largest high-precision sample of polarization measurements of GRB prompt emissions; An X-ray transient powered by a magnetar formed by the binary NS merger was discovered from the Chandra deep survey data. These studies have given tight constraints on the equation of state for the NS and fundamental physics related with strong magnetic fields, etc. GECAM discovered many GRBs and SGRs, realized the real-time onboard trigger and alert technique and guided multiwavelength telescopes for joint observations.

In the field of fast radio bursts (FRBs): Insight-HXMT discovered and identified the X-ray counterpart of FRB 200428 from the galactic magnetar SGR J1935+2154; The observations using FAST have provided the most stringent upper limit of the radio flux from the magnetar, and find the low association rate between the X-ray bursts and FRBs from the magnetar, these FRB studies with Insight-HXMT and FAST were selected as one of the top ten scientific breakthroughs in 2020 by *Nature* and *Science*, "an FRB comes from a magnetar in the Milky Way"; FAST also found the diverse evolution patterns of the polarization angle in repeating FRB 180301, and revealed the bimodal distribution of burst energy in repeating FRB 121102; Several physical origin models (an asteroid and asteroid belt accreted by an NS, the spiral of binary NS, a white dwarf accreted by an NS, and the collapse of the crust of a quark star, etc.) and radiation mechanisms for FRBs were proposed; FRBs are used as tools to probe cosmology and test fundamental physics.

In the field of pulsars and NSs: FAST made great promotion in the large sample search of pulsars, found the evidence for three-dimensional spin-velocity alignment in a pulsar, and performed the high-precision timing measurement, laying a foundation for future scientific goals, e.g., testing theories of gravity; Using domestic and foreign telescopes, a series of research progresses have been made in nanohertz GW search, pulsar timing array, and pulsar radiation, etc.; The PolarLight project captured the change and recovery of X-ray polarization from the Crab Nebula after its sudden glitch; A transient low-magnetic magnetar was discovered in the south of supernova remnant Kes 79; Chinese scholar was invited to publish a review of interstellar and intergalactic magnetic fields on *ARA&A*.

In the field of X-ray binaries: The mass of the black hole in an ultra-luminous X-ray source was first measured; A relativistic baryon jet was first discovered in the ultra-soft X-ray source; Insight-HXMT directly measured the strongest magnetic field so far in the universe, discovered a relativistic jet closest to the black hole, and a high-speed outward-moving plasma stream from a black hole; A method using millihertz quasi-periodic oscillation to constrain the NS radius was proposed, and gave a lower limit of 11 kilometers for the NS in 4U1636-53.

In the field of active galactic nuclei (AGN): A rapid inflow as the accretion fuel was directly observed using the Doppler shift of the hydrogen and helium absorption lines; The long-standing puzzle "the origin of the broad-line regions in AGNs" has been solved; A new method to measure cosmological distances of AGNs has been proposed; The wind near the AGN accretion flow was systematically studied, from which the author was invited to write a review paper on black hole accretion for *ARA&A*; 1,392 blazars from the Fermi data were classified by Bayesian method, and their radiation mechanism was discussed; The quasi-periodic oscillation of monthly magnitude was first found for blazar PKS 2247-131, providing direct observational evidence for the spiral magnetic field in the jet; The analytical solution of the radial equilibrium equation for a Poynting-flux dominated jet was obtained; Chinese scholars participated in the international collaboration of black hole imaging with Event Horizon Telescopes.

In the field of high-energy cosmic rays and gamma-rays: The DAMPE accurately measured the energy spectra of cosmic positrons, electrons, protons, and helium, discovered the new spectral structure, and obtained the gamma-ray spectra with the most sensitive survey; The Tibet ASγ experiment discovered a candidate source for accelerating ultra-high-energy cosmic rays, and first found evidence for ubiquitous Galactic cosmic rays beyond PeV energies; LHAASO discovered a dozen of PeVatrons and the highest-energy gamma-ray, and measured the UHE gamma-ray brightness of the standard candle Crab Nebula, launching ultra-high-energy (UHE) gamma-way astronomy era; The first ultra-bright infrared galaxy with gamma-ray emission was found from the Fermi data, extending the infrared-gamma-ray correlation of galaxies to the higher luminosity range; Using the Fermi-LAT data, it was found that gamma-ray emissions of galaxy M31 concentrate in the center of the galaxy, rather than along the disk.

In the field of high-energy neutrinos: The multi-emission-zone model was proposed to interpret the origin of high-energy neutrinos associated with blazars; The model of high-energy neutrinos from tidal disruption events (TDEs) of supermassive black holes was proposed, explaining the association between the high-energy neutrinos and TDE AT2019dsg; The fraction of protons in the GRB jet was constrained according to the neutrino observation from IceCube.

In the forthcoming years, the Advanced Space-based Solar Observatory (ASO-S), the Einstein Probe (EP), the Space Variable Objects Monitor (SVOM), and the Chinese Survey Space Telescope (CSST) will be launched during the 14th Five-Year Plan period. China has officially become a founding member of the international organization of Square Kilometer Array (SKA), and SKA-1 will be accomplished before 2025. Several 2-meter-level aperture optical time-domain survey telescopes under the cooperation between domestic universities and Chinese Academy of Sciences will be completed during the 14th Five-Year Plan period. The research of space astronomy and high-energy astrophysics in China will continue to maintain a high-speed development trend.

Celestial Mechanics and Astrodynamics

Zhou Liyong, et al.

In the past decade, the rapid progress in the observations of the Solar system and extra-Solar planetary systems and the implementation of deep space exploration programs have broadened the fields and deepened the theoretical researches of celestial mechanics and astrodynamics.

1. Fundamental theory of celestial mechanics

Researchers from Shanghai Astronomical Observatory (SAO) and Nanjing University (NJU) worked on the expansion of the disturbing function. Two new expansions that are both globally convergent except for collision singularities have been established. And these expansions are widely applicable including in the case of crossed orbits, and can be applied to the analyses of long-term orbital dynamics for near-Earth objects (NEOs), comets, Plutinos, and Trojans. The new expansion has also been used to obtain the structures of the phase space of mean motion resonances (MMRs) for highly inclined or even retrograde orbits. A systematic theoretical study on the orbital stabilities of the unrestricted three-body problem has been carried out, and the parameter spaces for special dynamical phenomena, such as orbital reverse and orbital exchange, have been investigated. An interior tidal dissipation model applicable to arbitrary rotation orientation has been developed, to show the dynamics of planetary orbital migration due to the cooperation between such tidal dissipation and the third-body gravitational perturbation.

2. Dynamics of small celestial bodies in the Solar System

NJU and Purple Mountain Observatory (PMO) carried out systematic investigations on the motions of Kuiper Belt objects (KBOs) and Trojan asteroids. The dynamics of Kuiper belt objects with high orbital inclinations in several resonances has been analyzed in depth, and the migration of Neptune and the capture of these resonant KBOs are reproduced in simulations. Consequently, the promising region of discovering these objects in future observations is predicted, and more constraints on the migration model of planets are revealed.

The systematic study of the dynamics of Trojan asteroids completely solves the problem of orbital stability of such objects in the current configuration of the Solar System, and provides a complete description of the dynamical mechanisms that influence the motions of Trojan asteroids of each planet. The regions and probabilities of finding the potential Trojan asteroids of Neptune and Uranus are predicted, and the possibility of the existence of undiscovered primordial Trojan asteroids of Earth and Venus is rejected.

The important influence of dwarf planets in the Kuiper Belt has been analyzed by calculating the total orbital angular momentum of all small celestial bodies in the solar system, and from these calculations, the constraints on the mass and orbital parameters of the putative Planet Nine have been raised.

3. Extra-Solar planetary systems

In NJU, PMO, and SAO, the traditional studies on celestial mechanics have been extended to the area of extra-Solar planetary (exoplanetary) systems.

For planetary systems orbiting a single star: A scenario has been proposed to explain the relation between the circumstellar dust disks and atmospheric metal pollution of white dwarfs, in which the asteroids in the disk are perturbed by the giant planets and falling into white dwarfs; It is revealed that the basic features of the first-order MMRs can be reproduced from the time-series astrometric data of Gaia satellite; The analytical theories of celestial mechanics have been widely developed and applied to confirm the exoplanets from the candidates, and to obtain the statistical properties of exoplanetary systems.

For planetary systems in binary star system: A scenario of inward migration of terrestrial planets through planetesimal scattering is proposed; The conditions for the occurrence of “close transits” of exoplanets are described, and as a result of the close transits, the increase of transiting time is beneficial for the accurate measurement of exoplanetary atmosphere.

For planetary systems in star clusters: The

multiplanetary systems in open clusters are found fairly stable, and most planetary systems that orbiting a single star can maintain their original configurations even if the hosting star is scattered out of the cluster; The photoevaporation effect of massive stars in clusters on the lifetime of protoplanetary disks is analyzed; The correlation between the occurrence rate of planetary systems and the dispersion velocity of stars in the clusters is obtained statistically, which serves as the evidence for the correlation between the planetary occurrence rate and the evolution history of stars in clusters.

4. Astronomical dynamics

Many efforts from SAO and others have been devoted to the development of BeiDou Navigation Satellite System (BDS). They are deeply involved in the development of information processing system for projects of BDS-2, BeiDou Experiment Support System, and BDS-3, designing and implementing the first real-time information processing system for massive multi-type data in China.

A precise orbit determination processing method for GEO satellites based on the constraints of bidirectional time alignment between satellite and ground is established, which breaks through the problem of high-precision real-time orbital determination of GEO satellites. Based on the BDS intersatellite link data, the satellite-ground and satellite-satellite joint method for precise orbit determination and time synchronization is developed, breaking through the difficulties arising from high-precision orbital determination of MEO satellites and clock error by the ground-based regional monitoring network, and realizing the high-accuracy service of navigation, positioning and timing by of BDS-3 to global users.

SAO, PMO, NJU, and other institutes have also been deeply involved in space missions of exploring the Moon and Mars. Based on the high-precision VLBI measurement, the S/X-band ΔDOR measurement technology and the multi-detector same-beam VLBI observation technology have been developed, and the measurement accuracy has reached 0.2ns and ps level, respectively. By combining the distance ranging, velocity measurements, and VLBI data, the measurement accuracy for lunar and deep space probes reaches 20m for circumlunar orbit determination, 100m for lunar surface positioning, and meter-level for relative positioning of the rover, while it's better than 2km and 100m for Earth-Mars transfer and circum-Mars orbit determination, respectively.

NJU with other institutes have constructed the dynamical alternative orbits near the equilibrium points of the Earth-Moon system, and systematically studied the types of Earth-Moon transfer orbits; The trajectories for spacecraft to orbit around, fly by, and land on small irregularly shaped celestial bodies have been analyzed and designed; The spin-orbit resonances in binary asteroid systems are systematically studied, and a method for fast calculation of the spin-orbit coupling is proposed.

The autonomous orbit determination algorithm is improved. A centralized autonomous orbit determination strategy for satellite constellations is proposed, which implements the autonomous orbit determination based on intersatellite links. The algorithm for orbit determination and cataloging for space debris is improved to include complex perturbations and be applicable to various orbit types. Experts from these institutes have participated in the development of NEOs alert system, composing the software for orbit prediction, initial and precise orbit determination, impact risk assessment, and active defense strategy for hazardous NEOs.

5. Astronomical ephemeris

PMO made important achievements in orbit fitting, stellar parametric determination and empirical mass-luminosity relations in binary and triple stars system, high-precision ephemeris and mass determination of asteroids, natural satellites ephemeris, and gravitational field parameter determination, and also in numerical ephemeris for planets and evolution of major fundamental frequencies in the Solar System. The standard astronomical almanacs in China are provided and maintained. Multiple kinds of astronomical almanacs, such as *Chinese Astronomical Almanac*, are regularly published. They also developed astronomical navigation software for the Navy, and provided almanac data for the National Marine Data and Information Service, Tiananmen Administration Committee, and other departments, and the internet almanac service for the general public. They formulated the national standard for Calculation and promulgation of the Chinese calendar. The "high-precision optical observation platform for Solar System celestial bodies (International number O49)" has been established, and the obtained data play an important role in the theoretical study of the motion of Solar System celestial bodies.

6. Relativistic celestial mechanics

Shanghai University of Engineering Science and Nanchang University designed explicit symplectic integrators for curved spacetimes, and developed some post-Newtonian (PN) theories, including the construction of canonical conjugate spin variables and coherent PN Lagrangian equations and the nonequivalence of Lagrangian and Hamiltonian approaches at same PN orders. PMO and NJU constructed a relativistic astronomical reference system which describes systems of celestial bodies on different scales, and investigated the second PN light propagation and its astrometric measurement in the solar system. SHAO gave analytical solutions of orbital motion for the extreme-mass-ratio relativistic two-body problem in the effective-one-body framework, constructed a time-frequency transfer relativistic model and relativistic geoid, and studied the propagation of light near slow-moving celestial bodies.

History of Astronomy

Sun Xiaochun, Wang Guangchao

As a major part of the history of science and technology, the history of astronomy has made a series of important progresses in the past decade, which is embodied in three aspects: project research, academic exchange, and international influence.

The research on the history of Chinese astronomy has maintained a close relationship with astronomy. The research issues have always grasped the scientific problems in the history of astronomy, from cosmology to astronomical records, from astronomical observation to astronomical instruments, from astronomical theory to calendar calculation, from star catalogue to star map, from Chinese astronomical tradition to astronomical exchanges between China and foreign countries, all of which have been continuously supported by the astronomical community. Over the past decade, the research on the history of astronomy has received the support of more than 20 general projects from the National Natural Science Foundation of China, which enables the research on the history of astronomy to accumulate and develop continuously.

At the same time, the history of astronomy has also made important breakthroughs in the field of Chinese Social Sciences. In the past decade, Chinese researchers in the history of astronomy have won three major projects from the China Social Science Foundation. "The research on the general history of the Chinese calendar", presided over by Northwestern University, studied the "restoration" of the algorithm of the ancient Chinese calendar, which makes the algorithm logic of the ancient Chinese calendar more distinct, and shows the scientific nature of the ancient Chinese calendar. "The collation and comprehensive study of ancient astronomical maps of China, Japan, and South Korea", presided over by University of science and technology of China, collected, sorted, and digitized Chinese ancient star maps scattered in China, Japan, South Korea, North Korea, Europe, and North America, and made an important contribution to the study of astronomical exchanges among East

Asian countries. "The study of astronomy spread along the Silk Road in the Han and Tang Dynasties", presided over by University of science and technology of China, verified and combed the knowledge of astronomy from outside China and calendars introduced into China in the Han and Tang Dynasties, studied their influences on Chinese astronomy and calendar, and presented a panoramic picture of Sino-foreign astronomy exchange in the Han and Tang Dynasties, which is both macro and comprehensive and without lack of details. The establishment of these major projects indicates that the study of the history of astronomy has also been widely recognized in the field of history, making it a truly interdisciplinary subject of science and humanities.

The organization of the history of astronomy project has promoted the research of the history of astronomy, and the research results are remarkable. More than 5 kinds of works have been published in recent ten years. Among them, *Textual Research on the Records of Celestial Phenomena in Various Histories* (2015) uses modern astronomical calculation methods to textual research, and tests all the records of celestial phenomena in the *Twenty-Four Histories*, the "Benji" and "Astronomical Chronicles" of the draft in the history of the Qing Dynasty; *Foreign Astronomy of the Tang Dynasty* (2019) examines the interactive relationship between foreign astronomy and social, religious, political, cultural, and other factors of the Tang Dynasty; *A Study of Heaven – the Spread of Jesuits and Astronomy in China* (2019), with Jesuits and astronomy as the theme, systematically expounds the aspects of Catholic missionaries and the introduction of European astronomy into China from the perspective of global history and cross culture; *Research on the Datong Calendar* (2020) systematically introduces various works related to the Datong calendar on the basis of a large number of new historical materials, and reveals the compilation, use, and dissemination of the Datong calendar in the Ming Dynasty; *Chongzhen Astronomical Treatises* (2017) is the collation and study of important raw materials for the study of literary exchanges between China and the West in the late Ming Dynasty. In addition, researchers in the history of astronomy in China have published more than 40 important academic papers in academic journals at home and abroad.

The history of astronomy in China pays special attention to academic exchanges. Since 2017, the Chinese Society for the History of Science and Technology has held an annual academic meeting. The Special Committee on the History of Astronomy has held a special session on the history of astronomy at each annual meeting, with more than 30 scholars participating each time, demonstrating a very active academic community. In addition, China's research on the history of astronomy also has a good image in the world. In August 2012, the 28th General Assembly of the International Astronomical Union (IAU) was held in Beijing. The Special Committee on the History of Astronomy organized the exhibition of achievements in ancient Chinese astronomy, and a special science popularization report "Ancient Chinese Astronomy" which not only publicized ancient Chinese astronomy, but also further promoted the study of the history of astronomy. At the subsequent IAU congresses, Chinese scholars made invited reports. In August 2020, at the 26th International Conference on the history of science and technology held in Prague, Chinese researchers in the history of astronomy organized several symposia.

In the past decade, the research on the history of astronomy in China has become more and more influential in the world. In 2015, during the 29th IAU conference, a Chinese scholar was elected chairman of the IAU Scientific Commission on the History of Astronomy. In 2017, another Chinese scholar was elected chairman of the Ancient and Medieval Astronomy Commission of the International Society for the History of Science and Technology (IUPHST/DHST). These positions were hold by Chinese scholars the first time in the history of these organizations. In addition, many scholars in the history of astronomy were elected Effective Member or Corresponding Members of the International Academy of the History of Science (IAHS), reflecting the ever increasing recognition of Chinese scholars in the International academic community.

Astronomical Education

Li Xiangdong

The last decade is the fastest growing decade of astronomical education in China. By the end of 2021, 22 Chinese mainland universities had conducted astronomy education and research, 12 of which had established undergraduate degrees in astronomy. Table 1 summarizes the new astronomy majors or departments added since 2012.

Table 1: Construction of Astronomy Majors in China mainland universities since 2012

No	Astronomy Department or School	Year of establishment of the undergraduate major in astronomy	Year of establishment of the department
1	Department of Astronomy, Xiamen University	2012	2012
2	Department of Space Science and Astronomy, Hebei Normal University	/	2012
3	Department of Astronomy, Yunnan University	2014	2013
4	Department of Astronomy, West China Normal University	2015	2016
5	Department of Physics and Electronic Science, Qiannan Normal University for Nationalities	2015	/

No	Astronomy Department or School	Year of establishment of the undergraduate major in astronomy	Year of establishment of the department
6	School of Astronomy and Space Sciences, University of Chinese Academy of Sciences	2016	2015
7	School of Physics and Astronomy, Sun Yat-sen University	2017	2015
8	Department of Astronomy, Shanghai Jiaotong University	2017	2017
9	Department of Astronomy, Guizhou Normal University	2017	2019
10	Department of Astronomy, Tsinghua University	/	2019
11	Department of Astronomy, Huazhong University of Science and Technology	/	2019
12	Department of Astronomy, Guangzhou University.	/	2020

In addition, many universities have established master's or doctoral degrees in astronomy/astrophysics, such as Guangxi University, Hunan Normal University, Central China Normal University, Nanchang University, Nanjing Normal University, and Shandong University (Weihai), Shaanxi Normal University, Shanghai Normal University, Tianjin Normal University, Wuhan University, Yunnan Normal University, and Zhongnan University.

With the growing power of astronomy education in China, the Ministry of Education established the Steering Committee for the Teaching of Astronomy in 2013, and since most of the members overlap with the members of the Education Working Committee, the work of astronomy education in the past decade has been carried out jointly by the two committees.

The national standard for the quality of undergraduate professional teaching is the basic requirement that the professional class personnel training, professional construction, etc. should meet. It will be mainly used in three aspects: first, as a professional reference for setting up specialty; second, as a guide for personnel training and professional construction; and third, as a reference for quality evaluation. The National Standard for the Quality of Teaching in Astronomy was first developed by Nanjing University in 2013, and lasted more than 4 years, during which many seminars and revisions were organized. The National Standard for the Quality of Undergraduate

Professional Teaching in General Higher Education, including this standard, is published by the Higher Education Press in 2018.

In accordance with the Basic Conditions Indicators for The Basic Conditions of General Higher Education (Trial) and the National Standard for the Quality of Undergraduate Professional Teaching in General Higher Education, Nanjing University took the lead in completing the development of the Standard for professional certification in Astronomy (Level 1) in 2018, that is, the basic requirements of the State for the running of astronomy majors in general higher education institutions.

In 2019, the General Office of the Ministry of Education issued a notice on the implementation of the "Double Million Plan" for the construction of first-class undergraduate majors, which plans to build about 10,000 first-class undergraduate majors at the national level, and 10,000 or so first-class undergraduate majors at the provincial level in 2019-2021. Nanjing University, University of Science and Technology of China, Guizhou Normal University, Peking University, and Beijing Normal University were selected as first-class undergraduate professional points at the national level.

The "Basic subject top student training pilot program" is a talent training program set up by the Ministry of Education in response to the "Qian Xuesen's question", aiming at training China's own academic masters. In 2018, the program was implemented in version 2.0, adding majors such as astronomy. Nanjing University, University of Science and Technology of China, and Peking University have been selected as the top student training program 2.0 bases for basic disciplines in 2019, 2020, and 2021, respectively.

In October 2019, the Ministry of Education issued the Implementation Opinion of the Ministry of Education on the Construction of First-Class Undergraduate Courses, which plans to build about 10,000 first-class undergraduate courses at the national level and 10,000 first-class undergraduate courses at the provincial level in about three years. Nanjing University's five major courses were selected as first-class undergraduate courses at the national level, and one course was selected as a demonstration course in the Ministry of Education' curriculum thinking and politics. In addition, many colleges and universities offered general and professional courses in astronomy on the platform of muse courses at home and abroad, which provides a wealth of learning resources for the students of universities.

We have held and participated in the organization of a number of educational and teaching seminars, and the national astronomical public election course research, which have effectively promoted exchanges and cooperation between universities in personnel training programs, curriculum, and teaching methods.

Astronomical Popularization

Cao Chen, Zhu Jin

Chinese National Astronomy Olympiad (CNAO) hosted by the Chinese Astronomy Society (CAS) and organized by Beijing Planetarium (BJP) and Popularization Working Committee (PWC)-CAS is successfully performed every year in two stages of Preliminary and Final Parts. Medals of Golden, Silver, and Bronze, and Best Performance Prizes are awarded in the Final Stages from three rounds of theoretical, practical, and observational, with a workshop on astronomy teachers during the period. After training at the qualification camp, participants of the national teams for three different international astronomy Olympiads are selected.

Chinese Undergraduate Astronomical Innovation Contests (CAIC) found by PWC-CAS in 2017 were held in 4 times between 2017-2021, hosted by CAS and organized by PWC-CAS, Education Working Committee of CAS, and local institutions. Both systems of CNAO and CAIC devoted to inspire and enhance the interest of astronomy for young students, and reserved outstanding personnel for Chinese astronomical research, education, and popularization careers.

Astronomy Teacher Training for Pingtang County hosted by Pingtang County Government and PWC-CAS and organized by BJP and Tianjin Astronomy Society etc. was performed in August 2017 in Pingtang County, Guizhou Province where the FAST radio telescope locates. 45 astronomy teachers from 7 elementary and middle schools attended, and astronomy courses were started as school courses spreading to about 4,000 students in 2017, and gradually extended to nearly 60 local schools in 2020. PWC-CAS hosted several National Science and Technology Teacher Training on Astronomy and Forum on Astronomy Popularization and Education (e.g., in Yanqing, Beijing in Sept. 2016, in Guizhou in Oct. 2017), and Astronomy Professional Training and Workshops in Urumqi, Xinjiang in May 2016.

The “IAU 100” series astronomical popularization activities with the theme of “Under One

Sky" celebrating the 100 anniversary of International Astronomical Union (IAU) were performed through the whole year of 2019, hosted by CAS and organized by PWC-CAS with supports from many domestic organizations and institutions. The main activities include the 2019 Chinese IAU100 opening ceremony held in BJP on Jan. 12 2019; the Chinese activity for IAU100 Global Project NameExoWorlds which was made during May-Nov. 2019 with procedures of public nomination, expert review, and public vote, and final confirmation from IAU, resulting the first exoplanet HD173416b discovered by Chinese astronomers being officially named with its parent star as Wangshu and Xihe.

PWC-CAS holds PWC-CAS Business Meeting every year and involved with organizing the parallel session on History of Astronomy, Education, and Popularization during the Annual Academic Conferences of CAS. During the Forum on Astronomy Popularization Volunteer Service Operation held on May 29-31 2019 in Hebei Normal University, PWC-CAS hosted the opening ceremony for CAS astronomy popularization volunteer service operation, and the photographic exhibition on Elegant Demeanour of Chinese Astronomers.

PWC-CAS encourages all CAS members to organize all kinds of astronomy popularization activities, to participate in National Science and Technology Week and National Scientific Popularization Day and other events, and to explore the new trend and new idea of astronomy popularization including online popularization activities under influences from the covid-19 situation and the 'double reduction' policy.

Astronomical Terminology

Yu Heng

In August 2013, *the Cross-Straits Astronomical Terminology* was officially published by Science Press. A modest inauguration ceremony was held at the third meeting of the Sixth Astronomical Terminology Committee in October of the same year. This work started in 2004 and finally came out after nearly 10 years of continuous efforts by colleagues in the astronomical community on both sides of the Taiwan Strait. This is the result of enhanced exchanges between astronomers from both sides of the Taiwan Strait and the initial achievement of the "Cross-Straits Astronomical Terminology Working Committee". It has facilitated academic communication, scientific cooperation, and popular education of astronomy across the Taiwan Strait. It laid a solid foundation for further consistency in astronomical terminology across the Taiwan Strait in the future.

With the continuous improvement of the nomenclature database and the addition of new terms, the difference between the content of the website and the original paperback astronomical dictionary has become more and more significant. The need for a new edition of the paperback astronomical nomenclature dictionary has become more and more urgent. Thanks to the efforts of many members, the new edition of *the English-Chinese Glossary of Astronomical Terms* was finally supported by the Sanxia Science and Technology Publication Grant Scheme of the Chinese Association for Science and Technology, and published by the China Science and Technology Press in December 2015. The new edition of the Dictionary of Astronomical Terms is automatically typeset by a computer program. It is fully consistent with the database and completely avoids the errors caused by manual entry and typesetting. The new edition contains about 26,700 entries. Compared with the previous edition published in 2000, the new edition contains about 6,000 new entries and 3,000 new notes; more than 2,100 typographical and input errors have been corrected, case and hyphen usage has been standardized, and irregular terms have been removed.

The new dictionary was distributed as a conference material at the annual astronomy conference in the same year. This greatly facilitated domestic astronomy students and researchers, and expanded the influence of the terminology work.

After the publication of the new edition of the Astronomical Dictionary, the work of the Astronomical Terminology Committee focused on the third edition of *Astronomical Nomenclature (Definition Edition)*. As the first committee to publish a dictionary with definitions of disciplinary terms, the Astronomical Terminology Committee plans to double the total number of entries to 4,500, based on the original 1998 edition, in order to fully reflect the rapid development and latest achievements in astronomy. The compilation of the new Definitions Edition is divided among 10 sub-disciplines. After several years of intensive work, the total number of entries in the new edition of *the Definitions of Astronomical Terms version* actually reached 4,873, exceeding the intended target by nearly 400. The trial version was released to the community in September 2021, and is expected to be officially published in 2022.

Besides the dictionary compilation work, the Astronomical Terminology Committee has also been working hard to improve the quality of public services and user experience. In 2017, a new version of the Online Glossary of Astronomical Terms is launched. It replaced the old system that had been running for many years, and was incorporated into the China Virtual Observatory platform. The new version of the website has been significantly improved in many aspects, including security, reliability, adaptability, operational efficiency, and user-friendliness.

In addition, in order to help China's Mars exploration mission and serve the public, the Astronomical Terminology Committee has organized efforts to translate for the first time all 1950 (as of 15 July 2020) names of Mars topographic features released by the International Astronomical Union (IAU) into Chinese in 2020. After validation by experts in the relevant fields, all the data are open freely for the public. This provides a useful reference for activities such as research and science education in related disciplines. Moreover, the National Astronomical Science Data Centre has created a special 3D visualization page for this release, which can visually show the actual location of the Martian names on the Mars map, facilitating the identification and study of Martian place.

Astronomical Journal

Hou Jinliang, et al.

Research in Astronomy and Astrophysics (*RAA*) was renamed from *Chinese Journal of Astronomy and Astrophysics* (*CHJAA*) in 2009. It is the only SCI international astronomy journal (published monthly) in the field of Astronomy and Astrophysics in China. It has a high degree of internationalization and a certain degree of international influence.

In the past decade, *RAA* has actively served a number of large astronomical science projects, and has published a series of Special Issues for them, such as special issues for China's Lunar Exploration, LAMOST, FAST, Advanced Solar Observatory, and Site Selection in the Mainland of China. At the same time, *RAA* has also published a number of internationally influential papers, including the special issues of IAU conferences, the white paper for the Thirty-Meter Telescope (TMT), the 30th anniversary of the Hubble Space Telescope, and other important review articles such as exoplanets. During the 2012 IAU General Assembly, *RAA* hosted the official newspaper during the meeting period, and organized a series of scientific activities.

Since 2012, *RAA* has obtained some important Funding and National Awards. Encouraged by national policies, the number of *RAA* submissions and publications has increased substantially in recent years. Compared with 2012, the number of *RAA* submissions and publications has nearly doubled, but the overall quality of manuscripts is still far from the world's first-class astronomical journals.

Acta Astronomica Sinica has gone through more than 60 years of development, covering all fields of Astronomy and Astrophysics. Through the efforts of the editorial committee and the editorial office, considerable progress has been achieved during the past decade. The influence of journal is constantly increasing, and the average cycle of publication has been shortened. The content has been constantly enriched by adding new columns including doctoral dissertation abstracts, review articles, large scientific projects, astronomical education, as well as comment on hot issues. The submission platform was upgraded, and the WeChat public account for the journal was opened.

Progress in Astronomy was founded in 1983. In the past decade, the range of journal papers has evolved from mainly the review of the frontier fields of astronomy to now both review and research papers. At the same time, the scope of published papers is no longer limited to the fields of astronomical observation and technology. Articles on applications such as deep space exploration, global navigation and positioning are also included. In addition, the quality of the edit has been raised, and the overall quality has been excellent in the editing quality inspections organized by relevant departments of Shanghai and the Chinese Academy of Sciences.

Astronomical Research and Technology was officially published in 2004, after being renamed from *Journal of Yunnan Observatory* (founded in 1977). It mainly publishes academic papers related to technologies, methods, and instrumentations related to astronomy. The journal also reports new discoveries in astronomy. In 2015, the website of the journal was updated, and a new submission and review system was activated. All published papers since 1977 were uploaded and are free for downloading. The Journal has been included in various database systems, such as the Chinese Science Citation Database (CSCD), the International Astronomical Databases ADS and CDS. In 2022, the publication period of the Journal will be changed from quarterly to bimonthly.

Astronomical Library

Hou Jinliang, et al.

The National Astronomical Observatories (NAOs) Library of the Chinese Academy of Sciences (CAS) is located at the headquarters of the NAOs, formerly known as the Beijing Astronomical Observatory Library. The NAO Library provides documentation and information services for all basic scientific activities of the NAOs. It supports the operation of national big scientific equipment, such as: FAST, LAMOST, Tianwen-1, CSST, and so on. Currently the Library has a collection of 6,957 books, 2,795 journals related to astronomy and astrophysics.

The Purple Mountain Observatory (PMO) Library of the CAS can be traced back to 1935. It is one of the most abundant astronomical libraries in China and one of the largest professional astronomical libraries in East Asia. There are more than 300,000 books and periodicals (bound and single edition) in the library, including a variety of publications in the field of astronomy since its inception. It is one of the characteristic collections of the Chinese Academy of Sciences. In order to further promote the modernization of the library, PMO Library adjusts the resource structure configuration, enriches the collection resources, strengthens the sharing of resources, and strives to play an irreplaceable role for science research, personnel training, social services, and cultural exchanges. It tries to make continuous progress towards a modern library.

The Shanghai Astronomical Observatory (SHAO) Library of the CAS originated from the libraries of the Xujiahui and Sheshan Observatories founded by the French Catholic Church at the 19th century. It is the oldest library in China which collected a large number of astronomical books and Journals. Some professional astronomy books and periodicals can be dated back to more than one hundred years. At present, SHAO Library has a collection of more than 60,000 Chinese and foreign books. Among them, nearly 15,000 books have been digitized, and the Library also provides more than 90 pieces of various types of information products and 50 types of lectures every year.

The Yunnan Astronomical Observatory (YAO) Library of the CAS was established in 1972. The library has been committed to serving scientific research ever since. The library has shifted its focus to the protection of electronic resources with the popularization of network applications. The current collection resources include 7,921 English books, 561 conference essays, 1,205 reference books, 343 astronomical calendars, 9,737 Chinese books, and 2,011 Chinese periodicals.

The National Time Service Center (NTSC) Library of the CAS, originally named as the Library and Information Research Office of the Shaanxi Astronomical Observatory of the CAS, was established in 1966. The library currently has a collection of more than 76,000 books, including time-frequency, navigation, physics, astronomy, communication, and computer science. A relatively comprehensive collection includes the Proceedings for the annual academic meetings of the National Time and Frequency, China Satellite Navigation, as well as other conference proceedings, such as ION, PTTI, EFTF, and IFCS. There are a total of 118 kinds of database resources available.

The library of Xinjiang Astronomical Observatory (XAO) covers an area of about 120 square meters, with more than 8,000 professional astronomy books and more than 10,000 volumes of 38 journals. It is not only a necessary service platform for the scientific research of the XAO staff, but also a strong support for the literature services of the NAO. Since 2011, the library was added to the unified automation system of the CAS, and gradually formed special characteristic service capabilities of the XAO, and established a new document information service mode of the observatory.

With the rapid development of information technology, the libraries of the observatories of CAS are rapidly developing the construction of digital libraries in order to adapt to the changes in the methods of retrieval and acquisition as well as protection of literature resources. The digital libraries include tens of thousands of astrophysical and optical resources from Springer-Nature, Science, IOP, Elsevier, APS, IOP, SPIE, and many other publishing houses, and have formed a knowledge service center which can provide intelligence analysis for different research fields and research processes.

Astronomical Information

Cui Chenzhou, Li Shanshan

Compared with other committees, the Informatizaion Working Committee, Chinese Astronomical Society (IWCC) is undoubtedly young. On March 25, 2019, after the approved by the first meeting of the 14th Council of the Chinese Astronomical Society (CAS), first 42 members of the IWCC were completely released. The responsibilities of IWCC include: 1. Regularly organize academic activities; 2. Organize the formulation of technical and management standards of astronomical informatization; 3. Promote the outstanding achievements and advanced figures in informatization work; 4. On behalf of CAS, connect with relevant departments, institutions, and organizations of informatization worldwide; 5. Undertake affairs assigned by CAS.

On June 14, 2019, the founding meeting of IWCC with the information working meeting was held in Nanjing, the establishment of IWCC was announced, and the official website was unveiled. The members collectively reviewed the articles of association of IWCC, and joined 4 working voluntarily, including the scientific research informatization working group, management informatization working group, science and education informatization working group, and informatization infrastructure working group. The second and the third working meetings of the first committee of IWCC were held in Xiamen (2020) and Kunshan (2021), respectively. In addition to the special report and work summary, the meetings discussed the annual work and hot topics of the current astronomical informatics, and collective strength of IWCC to organize the annual meeting of Virtual Observatory and astronomical informatics, and the formulation of relevant standards and specifications for informatization, and the other related issues were fully discussed. The establishment of the information committee is a milestone in the development of astronomical informatics work in China.

On June 5, 2019, the general office of the

Ministry of science and Technology issued the notice, relying on the National Astronomical Observatories (NAOs), co-constructed by Purple Mountain Observatory, Shanghai Observatory, and the Computer Network Information Center, the National Astronomical Data Center (NADC) has been officially included in the list of national science and technology resource sharing service platforms. In October, 2019, the NAOs of Chinese Academy of Sciences, Lijiang municipal government, Yunnan Observatory of the Chinese Academy of Sciences, and Yunnan University signed a cooperation framework agreement to build the Lijiang branch of NADC. In March 2021, the signing ceremony of the sub center of NADC was held at the NAOs in Beijing, Tianjin University, and Guangzhou University have signed cooperation and co-construction agreements with NADC, which will undertake the construction tasks of the NADC technology research and development (R&D) and innovation center and the Greater Bay Area center of NADC, respectively. In December 2021, NADC signed a cooperation agreement with Central China Normal University, Hebei Normal University, and China West Normal University to jointly build an education R&D Application Center, coordinate the implementation of curriculum R&D and education and teaching practice based on real astronomical scientific data, and realize the deep integration of scientific data, information technology, and education and teaching content.

Since the establishment of the IWCC, all the members actively push the development of astronomical informatics in China. Since March 2019, seminars, technical training, and lectures have been held for many times. More than 10 thousand people participated in these activities in total. Various online and offline data-driven astronomical education and public outreach activities have been held for many times in Beijing, Shanghai, Liaoning, Xinjiang, Daqing, and other places, such as the official opening of Shanghai Planetarium in 2021; The WorldWide Telescope Guided Tour Contest has been held for five times; The series of activities of "Xinjiang Youth Science Popularization Tour" founded by Xinjiang Observatory have been held for more than 40 times. These activities and projects have a certain social influence, and help the education and public outreach in astronomy.

The purpose of IWCC is to unite and organize professionals in astronomical informatics related fields, carry out academic and technical exchanges and strategic research, formulate relevant standards, organize professional training, improve the R&D and application level of scientific research informatization and management informatization in astronomy study in China, and enhance international influence. In the future, IWCC will continue to work around this purpose and in combination with the current actual situation to continue to promote the development of astronomy and astronomical informatics of China from the aspects of astronomical scientific research, management, science and education, infrastructure and so on.

Young Astronomer

Tian Bin, An Tao

Chinese astronomical scientific and technological researchers are trending younger obviously. In view of this, in order to help young astronomers to build a good development environment, make them feel at ease in their research and promote their growth, in January 2016, young astronomers from various astronomical institutions of the Chinese Academy of Sciences and some universities initiated a proposal to the Chinese Astronomical Society for the establishment of a Youth Working Committee. The Council of the Chinese Astronomical Society studied and approved the establishment of the Youth Working Committee (YWC, when not specified below, 'Committee' refers exclusively to YWC) as one of the working committees under the Chinese Astronomical Society. The YWC is responsible for coordinating the training of young astronomers, and promoting academic exchanges and cross-disciplinary study among young astronomers.

In order to standardize and promote related work, the committee held a kick-off meeting in Beijing in November 2016. The meeting focused on the implementation regulations of the YWC, the organization of the "Youth Astronomy Forum", the planning of academic exchange activities for young Chinese astronomers, and the principles and ideas for young astronomers to participate in the scientific popularization work. To further expand the influence and representation of the YWC, the Committee held its second working meeting in Kunming, Yunnan Province in January 2018, focusing on the proposal of adding a certain number of young astronomers from universities to the Committee.

Since 2012, with the core group of the Youth Innovation Promotion Association (YIPA) of some astronomical institutes of the Chinese Academy of Sciences, young astronomers in China jointly established the "Youth Astronomy Forum", which was officially included in the series of annual conferences of the Chinese Astronomical Society in 2014. After the establishment of the YWC, the Forum has been hosted by the Committee, and has gradually become a brand platform for cooperation and exchange among young astronomers in China.

In order to further exploit the active role of young astronomers in the development strategy of astronomy in China, the YWC, in collaboration with relevant departments of the Chinese Academy of Sciences, organized the "Workshop for Young Astronomers" in Kunming, Yunnan Province, from 19 to 21 January 2018, which was hosted by the YIPA group of the Yunnan Astronomical Observatory and the Key Laboratory of Astronomical Structure and Evolution of the Chinese Academy of Sciences. The aim of the meeting was to listen to the ideas and suggestions of young astronomers in China on the development of astronomy in the 14th Five-Year Plan (2020-2025) and beyond, to start preparing for the 14th Five-Year Plan, and to discuss the key areas that Chinese astronomy should focus on. The conference discussed the frontier areas, the optimization of resource allocation, the construction, implementation, and operation of observational facilities. During the preparations for the conference, more than 10,000 suggestions were received through 15 sub-disciplinary WeChat groups, 60 suggestions were received by email, and 34 written reports were produced. 33 conference convenors were responsible for collecting and collating the suggestions, and 48 conference reports were produced, including 12 overview reports and 36 disciplinary development reports. More than 130 astronomers from 28 institutions in China attended the conference. The conference also opened a live webcast platform, and more than 600 people watched the conference through the live video platform. The conference provided a valuable channel for active young scientists to contribute their ideas to relevant large-scale scientific projects.

Since its establishment, the YWC has carried out a series of work in promoting the growth and communication of young astronomer community, and young scholars have improved their knowledge, learned from each other, and made progress together through the platform built by the Committee. Young scholars are the driving force of scientific research and the future of astronomy. In its follow-up work, the YWC will continue to build up the existing communication platform for young scholars, actively collect and understand the development problems and suggestions faced by young astronomers, formulate corresponding work schemes, and play a more active role in the academic success of young astronomers.

Female Astronomers

Wang Na

In August 2017, the Female Astronomers Committee was officially formed as one of the working committees of the Chinese Astronomical Society. Wang Na, a senior research fellow at the Xinjiang Astronomical Observatory (XAO), Chinese Academy of Sciences (CAS), served as the chief of the committee.

The main duties and responsibilities of the committee include: 1. Providing support for female astronomers to improve their academic abilities, and encouraging female astronomers to fully participate in the astronomical research and study. 2. Strengthening the publicity on the achievements of the female astronomers, improving their social status and positions in scientific and technological innovations, and actively inviting the female astronomers to join the committee. 3. Organizing academic conferences, supporting the female astronomers to extend their scientific collaboration, and providing them with a platform to present their talents and achievements. 4. Collecting and heeding their needs and difficulties in scientific research and development, and to submit relevant comments and suggestions to the attention of the relevant departments.

On 27 October 2018, the first symposium for female astronomers was held, and more than 70 experts attended. During the symposium, Yang Ji, a research fellow from the Purple Mountain Observatory (PMO), Chinese Academy of Sciences (CAS), delivered a speech, and read out the "Implementation for the Female Astronomers Committee"; Cui Xiangqun, Fellow of the CAS introduced the senior outstanding female astronomers in China; Wang Xiaobin, a research fellow from the Yunnan Observatories of CAS shared her research experiences on asteroid over the years; Liang Chunyan, a research fellow from National Astronomical Observatory of China (NAOC) introduced the work of the International Astronomical Union (IAU) Women in Astronomy (WiA) Working Group (WG).

All the participants expressed their views on

how to balance the work and family, and they all agreed that "all the female astronomers, especially the young astronomers with broad development space and resources, should champion the spirit of our senior female astronomers to inspire ourselves to overcome the difficulties and to pursue our dreams. We should uphold the women's typical characteristics of fortitude and determination to achieve a better career. We are convinced that in the near future more and more female astronomers will devote themselves to the cutting-edge scientific and technological researches in astronomy with their excellent results."

From 26-27 December 2019, the second symposium for female astronomers was successfully held at the Nanshan Station of XAO. More than 70 experts, researchers, and students from astronomical institutes and universities all over China attended the symposium. Wang Jingxiu, Member of the CAS at the National Astronomical Observatory (NAOC), objectively analyzed the opportunities and challenges that China astronomy is facing through comparing the astronomical literature data at home and overseas and using example contributions by female astronomers from all over the world. Cui Xiangqun, Member of the CAS at the Nanjing Institute of Astronomical Optics & Technology, introduced the current development of optical telescopes in China in the context of the international developing process for astronomical telescopes and cutting-edge technologies. She encouraged everyone to grasp the best development period. During the symposium, participants delivered reports on their works and achievements, showing gorgeous successes in the astrophysics, astronomical techniques, space targets and debris. Female astronomers are tireless in research work and in pursuing excellence, and they take the responsibility of scientists in the new era of China. Using their own efforts to hold up the beautiful starry sky, female astronomers have played an irreplaceable role in the development of astronomy in China.

The committee has suspended the symposium for two years due to the Covid-19. Despite such intense period, every female astronomer has been making their best contributions to the development of the astronomical science in China with their research abilities and talents.

Astronomical Society

Ning Zongjun

In 2022, the Chinese Astronomical Society will celebrate its centennial birthday, which is the almost same age as the Communist Party of China. According to the division of each decade as a stage, we must look back on the course of the Chinese Astronomical Society in the past decade (2012-2021). Under the leadership of the China Association for Science and Technology, relying on the strong support of member units of various groups, provincial and municipal societies, and the majority of members, facing the world's scientific and technological frontier and major national needs, the Chinese Astronomical Society has continuously explored, innovated, and developed in building major national projects, organizing and promoting academic exchanges, strengthening international cooperation, carrying out popular science publicity, and strengthening organizational construction, with remarkable achievements and fruitful results. Today, the Chinese Astronomical Society has become the most important social organization representing the national astronomical scientific and technological workers, to be the bridge and link of the party and government with scientific and technological workers.

1. Build a first-class academic exchange platform for astronomy in China

Over the past decade, the Chinese Astronomical Society has made its main task to hold various forms of academic exchange activities based on the development of related disciplines in the frontier of astronomy. The society strives to host national annual academic conferences once a year to build itself into a first-class domestic academic exchange conference in astronomy. At the same time, it also cooperates with various professional committees and working committees to jointly hold a number of domestic and foreign academic conferences. According to statistics, in the past decade, there have been totally 70 academic activities, more than 100,000 participants, and more than 4,000 conference reports and papers.

Over the past decade, the annual academic conferences of the Chinese Astronomical Society have been held seven times respectively in Suzhou, Xi'an, Urumqi, Kunming, Delingha, Beijing, Nanchong, and other places. Representatives from more than 100 units in China, with a total of nearly 9,000 people, have exchanged more than 2,600 reports, more than 100 specially invited reports, and 40 advanced popular science reports and advanced astronomy lectures for the public. In the past two years, influenced by COVID-19, the academic annual meetings in 2020 and 2021 were combined with online and offline.

2. Strengthen international cooperation and exchanges, enhance the international status of China's astronomy, and actively respond to IAU100 activities

The Chinese Astronomical Society and its member units actively carry out academic exchanges with the International Astronomical Union (IAU), actively recommend IAU members, and participate in international activities led by IAU. Over the past decade, the Chinese Astronomical Society has organized a delegation to participate in the 29th session (in Hawaii, the United States in 2015) and the 30th session (in Vienna, Austria in 2018). A total of more than 80 IAU members have been recommended. They are from the front line of domestic scientific research institutes and universities to engage in astronomical work. The next (31st) IAU conference is planned to be held in Korea in August 2022. One of the highlights of the 30th Congress is the launch of IAU100 project. 2019 marks the 100th anniversary of the International Astronomical Union (IAU100). To commemorate this landmark event, IAU will organize a year-round celebration to increase public awareness of the important impact of astronomy as a tool in education, development, and foreign exchange by publicizing astronomical discoveries over the past century. The theme of IAU100 is "under the same sky". Chinese mainland China's IAU100 launching ceremony hosted by the Chinese Astronomical Society in 2019 was held by the popularization work committee and the Beijing planetarium in Beijing planetarium on January 12, 2019, forenoon. IAU100 global celebration projects held in 2019 include: holding the "100 hours of astronomy" global stargazing activity from January 10 to 13; Under the banner of Einstein School, establishing a global campus network to study topics such as gravity and general relativity; On February

11, being IAU women's and girls' astronomical day, astronomical activities will be held on the United Nations International Women's and girls' science day to encourage more women, especially girls, to participate actively; Make use of existing education programs and form a new IAU night Ambassador network to make the concept of night protection deeply rooted in the hearts of the people; The roving exhibition introduces the important achievements of modern astronomy and space exploration through the progress of science, technology, and culture over the past century; The 50th anniversary of the lunar landing in July 2019; Actions to improve teachers' ability in scientific topics and teaching skills; The new "Exoworlds" competition will provide all countries in the world with the opportunity to name exoplanets; During the total solar eclipse on July 2, 2019, celebrate the centennial of solar eclipse observation and verification of Einstein's theory of relativity, etc.

3. Actively carry out popular science activities and astronomy competitions for middle school students

The aim of the Chinese Astronomical Society is to "popularize Popular Astronomy". Relying on the Popularization Work Committee, the Chinese Astronomical Society and astronomical societies, associations or astronomical groups, and group member units in various provinces and cities actively promote astronomical science popularization activities. According to incomplete statistics, from 2012 to 2021, there were more than 200 popular science lectures, with nearly 300,000 participants, 400 popular science publicities, more than 30 Youth Science and technology competitions, 150 youth winter and summer camps, and tens of thousands of participants. Typical popular science activities include the theme activity of astronomy popular science entering Wa mountain for the first time in 2014, the 8th astronomy popular science education forum in 2016, astronomy business training and teacher training class in 2017, the 4th National College Students' astronomy innovation competition from 2017 to 2021, the final of national middle school students' astronomy knowledge competition in 2019, the activity of "National Science and technology workers' Day" in 2020 In 2021, the 17th public science day with the theme of "micro world exploration Tour" and other activities. Founded in 2017, the Chinese undergraduate Astronomical Innovation Contest (CAIC competition for short) is the only National

Astronomy competition for college students in China. In addition, in July 2021, the Shanghai planetarium was completed and opened to the outside world, which is the largest planetarium in the world.

The Chinese Astronomical Society organized the National Astronomical Olympiad for middle school students for ten times respectively in Urumqi, Xinjiang, Kunming, Yunnan, Beijing, Weihai, Shandong, Taiyuan, Shanxi, Yining, Xinjiang, Shaoxing, Zhejiang, Guizhou and Langfang, Hebei, and organized Chinese teams respectively to Brazil, South Korea, Bangladesh, Lithuania, Indonesia, Romania, Russia, the Republic of Tatarstan, Bulgaria, India, Thailand, Sri Lanka, China (Beijing, Weihai, Lijiang), and Hungary participated in the International Astronomical Competition, and won 39 gold medals, 67 silver medals, and 26 bronze medals.

4. Institution construction of Chinese Astronomical Society

In the past ten years, the Chinese Astronomical Society has changed its membership twice through the 12th, 13th and 14th councils, and held the national member congress in 2014 and 2018, respectively. The Standing Council or the Council has been held for more than 50 times to jointly study and formulate the work and plan of the society. Implement the spirit of the 19th CPC National Congress, and strengthen the party building of the society. The Party committee of the Chinese Astronomical Society was established in 2016. In 2018, the Party committee was changed. More than 10 Party committee and expanded meetings were held, and more than 30 documents were collectively studied and discussed. The youth working committee was established on October 31, 2016, the female astronomer Committee was established on August 7, 2017, and the Informatizaion Working Committee was established on March 25 2019. So far, the Chinese Astronomical Society has 8 working committees and fund working committee, together with original 11 professional committees.

In order to commemorate Mr. Lu Tan promote the development of Chinese astronomy, encourage and commend young scholars who have made important and original achievements in astronomy and astrophysics research, the Sixth Executive Council of the 14th session of the Chinese Astronomical Society decided to set up the "Lu Tan Award" specially by the Chinese Astronomical Society, which will be awarded every two years, with no more than two winners each time. 2022 will be the first selection of Lu Tan award. The prize is supported by the asteroid foundation of Nanjing Purple Mountain Observatory.

A Chronicle of Major Astronomical Events (2012-2022)

2012 The 65-meter radio telescope (Tianma Telescope) in Shanghai was completed, becoming the largest single-aperture fully steerable radio telescope in Asia.

January 16-20, 2012 The International Telecommunication Union (ITU) 2012 Radio-communication Assembly (RA) was held in Geneva, Switzerland. Our proposal disapproved the current abolition of leap seconds, and maintained the current international standard time-the definition of Coordinated Universal Time (UTC) to remain unchanged, which won the support of most countries.

January 20, 2012 The Guizhou Provincial Institutional Establishment Committee Office officially approved the establishment of the Guizhou Radio Astronomy Observatory.

February 8, 2012 The 28th Chinese Antarctic expedition team finished their work. Nanjing Institute of Astronomical Optics and Technology had successfully mounted the first astronomical optical telescope AST3-1 with remote control and automatic tracking of our country.

February 22, 2012 The Tiananmen District Management Committee of the Beijing Municipal People's Government presented the national flag to the Purple Mountain Observatory, and the astronomical calendars researchers of Purple Mountain Observatory will provide the Tiananmen Square Management

Committee with the standard sunrise and sunset times daily for the following year.

April 11, 2012 A test-observation of Optical and Near-infrared Solar Chromosphere Explosion Telescope (ONSET) developed by Nanjing Institute of Astronomical Optics and Technology for Nanjing University had achieved the first solar image of full-disk and local image at IR band in China.

May 20, 2012 In Fuqing of Fujian Province, meteorite experts from both sides of the Taiwan Straits had authenticated a 585kg siderite, which was the heaviest private-owned siderite specimen in China.

May 21, 2012 An annular eclipse happened upon the southeast coast of China. Numbers of astrophile teams came to Fujian.

June 1, 2012 Xiamen University officially decided to re-establish the Department of Astronomy.

July 4-6, 2012 The 429th Xiangshan Science Conference with the theme of "New View of Astrophysics and Large-Aperture Radio Telescope" was held in Urumuqi.

July 15, 2012 The groundbreaking ceremony of the Xinjiang Qitai Astronomical Observation and Popular Science Education Base was held.

August 25-31, 2012 The 28th IAU General Assembly was successfully held in Beijing, which was the first time after China joined the IAU for over 70 years, and Xi Jinping had a talk at the opening ceremony as the national vice-president. The IAU ROAD-East Asia was established during the assembly.

August 31, 2012 Professor Liu Xiaowei of Peking University was elected as the vice president of the International Astronomical Union.

October, 2012 Tsingtao Observatory of Purple Mountain Observatory was awarded as "National Popular Science Base" from 2012 to 2016.

October 8, 2012 The payload of "Dark Matter Particle Detection Satellite" led by Purple Mountain Observatory successfully completed the beam test at CERN (European Centre for Nuclear Research).

October 15-16, 2012 "October Astronomy Forum - The Past, Present and Future of Chinese Astronomy - and the Celebration of Prof. Wang Shouguan's 90th Birthday" was held in Beijing.

October 26-27, 2012 The activity of "Astronomical Popularization into Temples, Medical Healthy into Tibet" was held in Shangri-La successfully

October 28, 2012 The inauguration ceremony of the Shanghai 65m Radio Telescope and the celebration of the 50th anniversary of the establishment of the Shanghai Astronomical Observatory and the 140th anniversary of the establishment of the station were grandly held at the site of the Shanghai 65m Radio Telescope in Sheshan, Songjiang.

November 19, 2012 The signing ceremony of the Memorandum of Understanding on Science and Technology Cooperation between the Purple Mountain Observatory and the Delft University of Technology (TU Delft) in the Netherlands was held in Nanjing.

November 26, 2012 The Re-establishing Ceremony of the Department of Astronomy of Xiamen University was conducted.

December, 2012 The Tianma Telescope joined the Chang'e 2 mission for the first time.

December 13, 2012 Chang'e-2 satellite successfully flew over the asteroid "God of War" Toutatis, and took the images of this asteroid for the first time in the international arena. This is China's first flyby detection of asteroids, but also the first international close detection of the "God of War".

2013 Tianma telescope had successfully participated in and completed VLBI real-time precision orbit determination for Chang'e-3 and subsequent lunar exploration missions and China's first Mars mission "Tianwen-1".

2013 Prof. Wang Jingxiu of National Astronomical Observatories has been selected as the academician of Chinese Academy of Sciences.

January 10-13, 2013 The 13th East Asia Regional Workshop on Submillimeter Receiver Technology was held in Nanjing.

January 22, 2013 The Chinese Spectral Radioheliograph (CSRH) of the National Astronomical Observatories successfully acquired its first radio image of the sun at 1.7 GHz.

March 27-29, 2013 The BigBOSS (predecessor

of DESI) collaboration meeting was held, and Department of Astronomy of Shanghai Jiao Tong University became the full institutional member of DESI (Dark Energy Spectroscopic Instrument) project in April 2018.

May 26-29, 2013 International cooperative meeting of Antarctic Survey Telescopes (AST3) was successfully held in Tengchong, Yunnan Province.

April, 2013 The key laboratory for planetary science was established in Shanghai Astronomical Observatory and Purple Mountain Observatory.

April 27, 2013 The key laboratory for lunar and deep space exploration of the Chinese Academy of Sciences was established and attached to the National Astronomy Observatories.

June, 2013 Joint Laboratory for Radio Astronomy Technology (JLRAT) Laboratories of National astronomy observatories joined the SKA Antenna (DISH) and Broadband Feed (WBSPF) working package consortium.

June 20, 2013 Prof. Xuano Haibe, President of the International Astronomical Union and President of the National Astronomical Observatory of Japan, was invited to visit Purple Mountain Observatory.

June 20, 2013 Prof. Xuano Haibe, President of the International Astronomical Union and President of the National Astronomical Observatory of Japan, was invited to visit Nanjing Institute of Astronomical Optics and Technology.

August, 2013 Fujian Astronomy Society had collaborated with Kaohsiung Astronomy Society in organizing Min-Tai Astronomy Youth Camp.

September 3-6, 2013 The 5th Cross-strait Astronomical Telescope and Instrument Symposium had been successfully held in Xishuangbanna, Yunnan.

September 9-13, 2013 The 7th International Symposium on Solar Polarization organized by Yunnan Astronomical Observatory was held in Kunming.

September 10, 2013 During the international symposium of “Science and Technology of Large Radio Telescopes” in Urumuqi, a memorandum of understanding for the scientific and technical cooperation in the field of radio astronomy about the project of 110m telescope at Qitai Xinjiang was signed by the Xinjiang Astronomical Observatory and the National Radio Observatory of USA.

September 25, 2013 Chinese Astronomical Society carried out the selection of the 4th Huang Shoushu prize, and Yang Xiaohu won this prize.

October, 2013 Academician Cui Xiangqun in Nanjing Institute of Astronomical Optics and Technology was awarded the 2013 “Ho Leung Ho Lee Foundation Science and Technology Progress Award”.

October 7, 2013 The South American Astronomical Research Center of the Chinese Academy of Sciences was established at the University of Chile.

October 7-12, 2013 The international conference “Energetic Particles in the Heliosphere” was held in Nanjing.

October 23, 2013 *The Cross-Strait Astronomical Terminology* was officially published by Science Press.

October 27-31, 2013 The 2013 annual academic meeting of Chinese Astronomical Society was held in Suzhou City. More than 500 participants attended. There were 10 specially invited talks, 2 high-level popular science talks, and 298 oral talks exchanged in 8 sub-meetings.

November 10, 2013 The 10th China-Australia Science and Technology Symposium on “Astronomy and Astrophysics” was opened in Nanjing. The conference was jointly hosted by the Chinese Academy of Sciences, the Australian Academy of Sciences and the Australian Academy of Technological Sciences and Engineering.

December, 2013 The “time.ac.cn” of National Time Service Center was rewarded “2013 Excellent Science Popularization Website of the Chinese Academy of Sciences”.

December 10-12, 2013 Nanjing Institute of Astronomical Optics and Technology successfully held the first “Antarctic Star” forum of young scientists.

December 14, 2013 The lunar landing was successful, and my country achieved the first soft landing on the lunar surface. The VLBI subsystem carried out the relative positioning of the "Jade Rabbit" lunar rover after the moon fell.

December 28, 2013 The Institute of Astronomy and Space Science of Sun Yat-sen University was unveiled at Zhuhai Campus, to rebuild the astronomical discipline.

December 31, 2013 The construction of Chinese Spectral Radioheliograph (CSRH) was completed and accepted.

2014 The Department of Astronomy, Xiamen University signed co-construction agreements with National Astronomical Observatories and with Shanghai Astronomical Observatories, and established the "SHAO-XMU Joint Center for Astrophysics".

January, 2014 The monitor of Antarctic seeing (DIMM) developed by Nanjing Institute of Astronomical Optics and Technology started to observe at the Taishan Station of South Pole.

January 22, 2014 The Large Synoptic Survey Telescope (LSST)-China Consortium signed a memorandum of understanding with LSST Corporation.

March 1, 2014 "LAMOST Galactic Study", a major national Science Foundation project, was started.

May, 2014 The fiber grating spectrograph developed by Nanjing Institute of Astronomical Optics and Technology for 2.4m telescope of Thailand passed the inspection that locally debugged and achieved the first light successfully in October of this year.

May, 2014 Tsingtao Observatory was named by Shandong Provincial Science Association and Shandong Provincial Finance Department as four-star Shandong Provincial Science Education Base.

May 6, 2014 TMT International Observatories Co., Ltd. (TIO LLC) was established, with members including university of California, California Institute of Technology, National Astronomical Observatory of Japan, National Astronomical Observatories of China, Indian Ministry of Science Technology, and Canadian Collegiate Union of Astronomical Researches.

June 23-27, 2014 The 22nd International Planetarium Society (IPS) Conference was held in Beijing.

July, 2014 The Extreme Adaptive Optics (Ex-AO) developed by Nanjing Institute of Astronomical Optics and Technology had successfully connected with the 3.6m New-Technology Telescope (NTT) of European South Observatory (ESO) as a guest device, and finished the test-observation satisfactorily.

July 17-18, 2014 Members of the JLRAT laboratory assisted the Ministry of Science and Technology to hold the S21 Xiangshan Science Conference with the theme of "China's Radio Astronomy Development and the Square Kilometre Array Telescope (SKA)" at the Xiangshan Hotel in Beijing.

July 22, 2014 The CarT-China-Argentina Radio Telescope project was officially approved by Argentina's Ministry of Science, Technology and Innovation (MinCyT).

August, 2014 "110m Large Aperture Fully Mobile Telescope (QTT) Key Technology Research" project (National Key Basic Research and Development Program) declared by Xinjiang Astronomical Observatory was successfully established.

August 21, 2014 The Key Laboratory of Space Astronomy and Technology of Chinese Academy of Sciences and the Key Laboratory of Computational Astrophysics of Chinese Academy of Sciences were established and attached to the National Astronomy Observatories.

September 11, 2014 Chinese Astronomical Society carried out the selection of the 12th Zhang Yuzhe prize, and He Xiangtao won this prize.

September 23, 2014 "National Astronomy Observatories - Guizhou Normal University Astronomy Research and Education Center & FAST Early Science Data Center" were inaugurated (in Guizhou Normal University).

October, 2014 Wang Yanan in Nanjing Institute of Astronomical Optics and Technology was awarded the 2014 "Ho Leung Ho Lee Foundation Science and Technology Progress Award".

October 15, 2014 The Indian solar chromosphere telescope had successfully mounted and debugged by Nanjing Institute of Astronomical Optics and Technology and Astronomical Instruments Co., Ltd.

October 27, 2014 2014 Academic Annual Meeting of Chinese Astronomical Society and the 13th National Member Representative Conference were grandly opened in Lintong District, Xi'an, Shaanxi Province. There were 131 representatives. The 13th executive council was elected and Wu Xiangping of National Astronomical Observatories became the chairman.

October 29, 2014 Purple Mountain Observatory was awarded the honorary title of "Outstanding Volunteer Unit" by the Ho Leung Ho Lee Foundation.

November, 2014 The translated *DK Illustrated Encyclopedia of the Universe* by Beijing Normal University won the Eighth Wu Dayou Award for Popular Science Writing, and was selected by the Ministry of Science and Technology as a National Outstanding Work of Science and Technology in 2016.

November 4-8, 2014 The 16th China International Industry Fair was held at the Shanghai National Convention Center. National Time Service Center's "Beidou High-precision Wide Area Augmentation Service System and User Terminal" project participated in the group exhibition of Chinese Academy of Sciences, and won Innovation Award.

December 15, 2014 The "Mid-Infrared Observation System for Accurate Measurement of Solar Magnetic Field (AIMS)" was approved by the National Natural Science Foundation of China.

2015 Prof. Jing Yipeng of Shanghai Jiaotong University has been selected as the academician of Chinese Academy of Sciences.

January, 2015 The Antarctic Sky Survey Telescope AST3-2 was successfully installed in Dome A, Antarctica.

January 7, 2015 The signing ceremony of the cooperation agreement between Yunnan Astronomical Observatory and Thailand National Astronomical Institute was held in the capital Bangkok of Thailand.

May 11, 2015 The first International GNSS (Global Navigation Satellite System) Space Signal Quality Monitoring and Assessment (SMAS) Seminar was held in Qujiang International Conference Center, Xi'an, China.

May 22, 2015 Beijing Planetarium was rated

as National Base of Scientific Popularization and Education (2015-2019).

June 8, 2015 The Space Debris Monitoring and Application Center of the China National Space Administration (CNSA) was inaugurated at the National Astronomical Observatories.

June 23, 2015 Purple Mountain Campus of Purple Mountain Observatory was renewed as a "National Popular Science Base" from 2015 to 2019 through a review.

August 6, 2015 The 500-meter Aperture Spherical Radio Telescope (FAST) Project Reflector Cable Network Manufacturing and Installation Project had passed the completion acceptance.

August 17, 2015 The Delingha Astronomical Science Center jointly built by the Purple Mountain Observatory and the Haixi Prefecture Government was officially opened.

August 30 - September 4, 2015 The 6th Cross-strait Seminar on Advanced Technology of Astronomy Telescope and Observation was successfully held in Hualien, Taiwan.

September 10, 2015 Chinese Astronomical Society carried out the selection of the 5th Huang Shoushu prize, and Wang Xiaofeng and Chen Xin won this prize.

September 16, 2015 The Institute of Physics and Astronomy of Sun Yat-sen University was established, which includes the Institute of Astronomy and Space Science.

September 22-24, 2015 Chinese Astronomical Society carried out the selection of the 1st Huang Runqian prize, and Jing Yipen and Wu Xuebing won this prize.

September 29, 2015 The launching ceremony of the name campaign for the Dark Matter Particle Explorer satellite was held at Purple Mountain Observatory. According to the results, the satellite was later named "Wukong".

October, 2015 The "International Joint Research Center for Radio Astronomy and Technology" declared by Xinjiang Astronomical Observatory was approved by the Ministry of Science and Technology and officially included in the sequence of national-level international scientific and technological cooperation bases.

October 19-21, 2015 The 2015 annual academic meeting of the Chinese Astronomical Society was held in Peking University. More than 800 participants attended. There were 8 specially invited talks, more than 300 oral talks and 30 posters in 8 sub-meetings.

November 22-25, 2015 The 16th East Asia Regional Workshop on Submillimeter Receiver Technology was held in Nanjing.

November 26, 2015 Miro Cerar, the prime minister of the Republic of Slovenia, visited Beijing Ancient Observatory.

December 1, 2015 The new edition of *English–Chinese Astronomical Terms* was published by China Science and Technology Press, with the support of the CCSA Sanxia Science and Technology Publication Grant Scheme.

December 14, 2015 "China SONG Telescope" successfully passed the site inspection.

December 17, 2015 Dark Matter Particle Explorer Satellite "Wukong" was successfully launched by the Long March 2D vehicle from the Jiuquan Satellite Launch Center, so that China finally achieved a zero breakthrough in astronomical satellites.

December 24, 2015 The first satellite of China's scientific satellite series, the Dark Matter Particle Explorer Satellite "Wukong", successfully acquired the first batch of scientific data on the seventh day after launch.

2016 Since 2016, a series of activities of "Xinjiang Youth Science Public Outreach Tour" have been held for more than 40 times of science lectures on astronomy, geography, and other disciplines.

2016 Research Professor Han Zhanwen of Yunnan Astronomical Observatory won "Ho Leung Ho Lee Foundation Science and Technology Progress Award".

2016 The Department of Astronomy, Xiamen University was authorized the Fujian Province Key Laboratory in Astrophysics and Astronomical Instruments.

January, 2016 Young astronomers from various astronomical institutions of the Chinese

Academy of Sciences and some universities initiated a proposal to the Chinese Astronomical Society for the establishment of a Youth Working Committee, which was studied and approved by the Council of the Chinese Astronomical Society.

January 1, 2016 China's launch of Dark Matter Particle Explorer Satellite was selected as one of the top ten news items in China's aerospace and the top ten news items in the world's aerospace.

April 11, 2016 Purple Mountain Observatory and University of Science and Technology of China (USTC) signed a framework agreement to jointly build a school of astronomy and space science for the integration of science and education.

April 12-15, 2016 The "27th International Symposium on Space Terahertz Technology (ISSTT)" was held in Nanjing. This conference is the first time that the ISSTT conference was held in China.

April 15, 2016 Malcolm Turnbull, the prime minister of Australia, visited Beijing Ancient Observatory.

May, 2016 The Xiangshan Scientific Conference on "Science and Technology Development Strategy of Chinese Ground-based Large Aperture Optical Infrared Telescope" was successfully held in Nanjing.

June, 2016 The modification of 1m telescope of MAO Observatory of Uzbekistan was successfully finished by Nanjing Institute of Astronomical Optics and Technology.

June 1-7, 2016 FAST Engineering participated in the National "Twelfth Five-Year" Science and Technology Innovation Achievement Exhibition; on June 3, Xi Jinping, Li Keqiang, Liu Yunshan, Wang Qishan and other party and state leaders went to the FAST booth to watch.

June 7, 2016 The China-US Universities Astronomy Collaboration Summit, with the theme "Our Universe, a World without Boundaries", was held in Beijing. The participants were invited to attend the Seventh round Conference of China-US High-Level Consultation on the People-to-People Exchange (CPE), and were received by Chinese Vice-Premier Liu Yandong and US Secretary of State John Kerry.

July 20-22, 2016 The 5th "Youth Astronomy

Forum" was held in Weihai, Shandong Province.

August 19, 2016 The wide-field photometric survey project of Yunnan Astronomical Observatory and Hong Kong Astronomical Society started to operate.

September 15, 2016 The "Tianji" gamma ray burst polarization detector was launched with the "Tiangong II" space laboratory.

September 20, 2016 Chinese Astronomical Society carried out the selection of the 13th Zhang Yuzhe prize, and Han Jinlin won this prize.

September 25, 2016 When FAST was put into operation, President Xi Jinping, general secretary of the Communist Party of China Central Committee and chairman of the Central Military Commission, sent a congratulatory letter.

October 15, 2016 Time, Frequency, and Satellite Navigation Academic Seminar and Celebration of the 50th Anniversary of the Establishment of National Time Service Center were held at National Time Service Center in Lintong, Xi'an.

October 17, 2016 The qualification review meeting of National Time Service Center Analysis Center of the International GNSS Monitoring and Evaluation System was held at the Xi'an National Civil Aerospace Industry Base. The review team agreed that National Time Service Center Analysis Center met the index requirements for the iGMAS analysis center access standards.

November 1-3, 2016 The 2016 annual academic meeting of Chinese Astronomical Society was held in Wuhan City. More than 750 participants attended. There were 8 specially invited talks, 337 oral talks and 40 posters in 8 sub-meetings.

December 5-7, 2016 "The Second Sino-Australian Astrophysics Symposium and Sino-Australian Astrophysics Seminar and China-Australia Astrophysics Symposium" was held in Suzhou, Jiangsu.

December 13, 2016 The world's first ultra-wideband (0.75~15THz) terahertz Fourier spectrometer operating in unattended working mode was successfully installed and obtained the long-period measured data of atmospheric transmittance in the terahertz to far-infrared spectrum at the Dome A

Observatory in Antarctica.

March 6, 2017 The Tsingtao Observatory of Purple Mountain Observatory was awarded the 2016 National Excellent Popular Science Base.

March 28, 2017 Purple Mountain Observatory was selected as the first batch of Ten Science and Technology Tourism Bases in China.

April 6, 2017 The establishment of the Department of Astronomy meeting was organized by Shanghai Jiao Tong University.

April 16-18, 2017 Chinese Astronomical Society carried out the selection of the 2nd Huang Runqian prize, and Zhao Gang and Dai Zigao won this prize.

May 4, 2017 The Kavli Institute for Astronomy and Astrophysics, Peking University (KIAA)'s 10th Anniversary Symposium was held at KIAA, Peking University.

May 10-12, 2017 The 6th "Youth Astronomy Forum" was held in Kedu Town, Pingtang County, Guizhou.

May 15-19, 2017 The 1st China-Europe Symposium on Solar Physics was held successfully in Kunming, Yunnan Province.

June 15, 2017 The first X-ray space astronomical satellite "Huiyan" was successfully launched.

June 16, 2017 National Space Science Center held a meeting on the comprehensive demonstration report of the Advanced Space-based Solar Observatory (ASO-S) satellite project, which has been qualified for the project and agreed to pass the review.

June 21, 2017 Chinese Astronomical Society carried out the selection of the 6th Huang Shoushu prize, and Dong Subo and Wang Huiyuan won this prize.

June 30, 2017 The national major scientific research instrument development project (recommended by the department) "Terahertz Superconducting Array Imaging System" successfully passed the project acceptance organized by the National Foundation of China.

July 6-9, 2017 "The Second Astronomical Silk

Road: International Symposium about Exchange of Astronomy and Astronomical Archaeology between China and Central Asia" was held in Urumuqi.

August, 2017 The Female Astronomers Committee was officially formed as one of the working committees of the Chinese Astronomical Society.

August, 2017 Fujian Astronomical Society had collaborated with Astronomy Club Union of Universities in Taiwan in organizing Fujian-Taiwan Undergraduate Astronomy Communication Camp.

August, 2017 Colleagues from Fujian, Taiwan, Hong Kong, and Macao had grouped together to celebrate 30th Anniversary of Chiayi Amateur Astronomers Association.

August 7-11, 2017 The annual academic conference of Chinese Astronomical Society was held in Urumuqi, Xinjiang. More than 800 participants attended. There were 8 specially invited talks, 359 oral talks and 40 posters in 8 sub-meetings.

August 18, 2017 AST3 collaborative team of the Chinese Antarctic Survey Telescope AST3 conducted effective observations of GW 170817 using the telescope AST3-2, which is operating at the Kunlun Station in the Chinese Antarctic, and detected the optical signal of this gravitational wave event.

September, 2017 The discipline of astronomy and space science at Nanjing University had been selected into the National "Double first class" Initiative.

September, 2017 The optical lattice cold atom strontium (87) optical clock (Strontium optical clock for short) developed by the research team led by the National Time Service Center successfully achieved closed-loop operation.

September 3-6, 2017 The 7th Cross-strait Astronomical Telescope and Instrument Symposium was successfully held in Yichun.

September 6, 2017 The second-stage satellite navigation test system based on transponder ranging, developed and constructed by National Time Service Center passed the acceptance test organized by the China Navigation Satellite System Management Office. The capacity of positioning, velocity, and timing (PVT) had a major breakthrough.

September 28, 2017 The seminar of "2017 Silk Road Astronomy and 60th anniversary of Xinjiang Astronomical Observatory" was held in Urumuqi.

October, 2017 A solar Multi-Conjugate Adaptive Optics (MCAO) experiment system was, for the first time, built and successfully obtained the first light based on 1-meter New Vacuum Solar Telescope in China, which promote China as the third country in the world to hold the technology of large FOV adaptive optics.

October, 2017 Unattended over-winter observation of Antarctic Survey Telescopes (AST 3-2) had successfully realized to contribute the detection of optical counterpart of gravitational wave.

October 10, 2017 The first results of the 500-meter Aperture Spherical Radio Telescope (FAST), including 6 new pulsars, were released under the auspices of the Chinese Academy of Sciences.

October 27, 2017 The overall acceptance of the "Shanghai 65m Radio Telescope System Development Project" was organized and completed in Shanghai.

October 27, 2017 The cooperated station of coronagraph was opened in Lijiang, Gaomeigu station, jointed by Japan National Astronomical Observatory, National Astronomical Observatories of China, and Yunnan Astronomical Observatory.

November, 2017 Han Zhanwen of Yunnan Astronomical Observatory was selected as the Academician of Chinese Academy of Sciences.

November, 2017 The translated *Cosmigraphics: Picturing Space Through Time* by Beijing Normal University was selected by the Ministry of Science and Technology as a National Outstanding Work of Science and Technology in 2017.

November 3, 2017 The Joint Center for Computational Astrophysics at the National Astronomical Observatories of China and KwaZulu-Knight University (NAOC-UKZN) in South Africa was established.

November 16, 2017 A new edition of the English-Chinese Astronomy Dictionary was officially launched and incorporated into the China Virtual Observatory platform.

November 24, 2017 The headquarters of Purple

Mountain Observatory was moved from Gulou to Xianlin Campus, and the unveiling ceremony of the bronze statue of Mr. Zhang Yuzhe was held in Xianlin Campus.

November 26-December 9, 2017 The training class of international young astronomers for the Maritime Silk Road was held successfully in Kunming.

November 27, 2017 A press conference on the first scientific results of the Dark Matter Particle Explorer satellite "Wukong" was held at the Chinese Academy of Sciences.

December, 2017 The organization signed a memorandum of cooperation on pulsar research (FAST- Fermi_LAT cooperation memorandum) with the Fermi LAT laboratory.

December, 2017 1.35m ultrathin non-spherical corrector successfully developed by Nanjing Institute of Astronomical Optics and Technology had helped the project ZTF of USA to achieve the first light.

2018 The Dark Matter Particle Explorer "Wukong" satellite, participates the "Celebration of the 40th anniversary of reform and opening up large-scale exhibition".

January, 2018 The key laboratory of FAST of CAS was established and located in NAOC.

February, 2018 The femtosecond optical frequency comb and its application research team of National Time Service Center successfully realized ultra-stable optical-cavity-based microwave source, which reached the international advanced level.

February, 2018 The second solar chromosphere telescope developed by Nanjing Institute of Astronomical Optics and Technology and Astronomical Instruments Co., Ltd. for Indian Institute of Astrophysics (IIAP) started to operate.

February 22, 2018 A newly discovered near-Earth asteroid (PHA) posing a potential threat to Earth, 2018 DH1, discovered by Purple Mountain Observatory China Near-Earth Object Survey Telescope on February 22, 2018, passed by Earth at a distance of 9.18 Earth-Moon distances.

March, 2018 The project of Wide Field Survey Telescope (WFST) jointly built by USTC and PMO was launched.

May 23, 2018 "High-precision Ground-based Timing System" proposal of National Time Service Center for the "13th Five-Year Plan" national major scientific and technological infrastructure was approved by National Development and Reform Commission.

June 21, 2018 Sun Yat-sen University formally approved to rebuild the Astronomy Department.

July 4, 2018 The launch meeting of the Advanced Space-based Solar Observatory (ASO-S) satellite was held in Beijing.

July 16-18, 2018 The first "First Cup" Future Technology Innovation Competition of the Chinese Academy of Sciences came to an end at Zijing Villa, Nanshan District, Shenzhen, and the "Miniaturized High-Performance Femtosecond Optical Frequency Comb" project of National Time Service Center won the prize of the competition.

August, 2018 Zhu Jin of Beijing Planetarium was elected to be a member of Executive Committee *WG* Small Bodies Nomenclature of IAU.

August 6-8, 2018 Chinese Astronomical Society carried out the selection of the 3rd Huang Runqian prize, and Li Yan won this prize.

August 14-17, 2018 7th "Youth Astronomy Forum" was held in Urumqi.

September, 2018 The middle-resolution spectrograph developed by Nanjing Institute of Astronomical Optics and Technology for LAMOST had successfully passed the local inspection, which opened the survey of LAMOST with the middle-resolution and achieved "Ten Progresses of Chinese Astronomy" in 2018.

September 13, 2018 The launch meeting of the "Frontier Research on Large Scientific Devices" project of the National Key R&D Program was held at Purple Mountain Observatory.

September 15, 2018 The first China Tianquan Lake Astronomical Forum opened in Xuyi.

September 27, 2018 Chinese Astronomical Society carried out the selection of the 14th Zhang Yuzhe prize, and Chang Jin of Purple Mountain Observatory won this prize.

October 27-31, 2018 Annual Meeting of Chinese Astronomical Society was held in Kunming, and 14th representative reference was held on October 31 to December 1 2017. More than 900 participants attended. There were 9 specially invited talks, 2 high-level popular science talks, 2 special lectures, 388 oral talks and 32 posters in 8 sub-meetings.

October 27, 2018 The first symposium for female astronomers was held.

October 29, 2018 The aurora project X-ray polarization measurement cube star (Polarlight) and the first validation test satellite of "Tiange Program", were successfully launched simultaneously.

October 30-31, 2018 The 14th national member congress of Chinese Astronomical Society was held in Kunming City. There were 128 representatives. The 14th executive council was elected and Jing Yi-peng of Shanghai Jiaotong University became the chairman.

November, 2018 GNSS space signal quality evaluation system with a 40-meter antenna was built by the signal quality evaluation team of National Time Service Center in Luonan County, Shaanxi Province, which adopted the world's first dedicated evaluation system realized GNSS satellite navigation signals.

November 5, 2018 The 13th United Nations International Committee on Global Navigation Satellite System (ICG) was opened in Xi'an, Shaanxi.

November 6, 2018 Prof. Chang Jin, a researcher of Purple Mountain Observatory, received the Ho Leung Ho Lee Foundation Award for Progress in Science and Technology.

2019 The Event Horizon Telescope (EHT) released the first-ever image of a black hole with SHAO leading the domestic research team.

2019 SHAO officially released the world's first SKA regional center prototype.

2019 The Department of Astronomy, Xiamen University joined the Chinese Astronomical Society as a group member.

2019 The Department of Astronomy, Xiamen University, renewed the co-construction and deepening cooperation agreements with Shanghai Astronomical Observatory.

January 7, 2019 Chinese Astronomical Society carried out the selection of the 7th Huang Shoushu prize, and Xie Jiwei of Nanjing University and Guo Jianhua of Purple Mountain Observatory won this prize.

February, 2019 For the kilometer-class satellite-ground bi-direction quantum entanglement distribution, the term of Chinese "Mozi" Quantum Satellite won the Newcomb Cleveland Prize by the American Association for the Advancement of Science of USA, as a participant, Xinjiang Astronomical Observatory won a medal of the Newcomb Cleveland Prize.

February 27, 2019 The research results of "Wukong" Dark Matter Particle Explorer Satellite (DAMPE) led by Purple Mountain Observatory of "First direct detection of the inflection of the electron cosmic ray energy-spectrum near 1 TeV" were selected as one of the 14th "Top Ten Advances in Chinese Science".

March 12, 2019 China and other seven countries formally signed the SKA Observatory Convention of the Square Kilometer Array.

March 19, 2019 Tsinghua University announced the establishment of the Department of Astronomy.

March 25, 2019 After the approved by the first meeting of the 14th Council of the Chinese Astronomical Society, first 42 members of the Informatization Working Committee were completely released.

April, 2019 A transient source driven by a magnetar formed by merger of a neutron star pair was firstly detected by USTC, which won the USTC President Prize for Outstanding Studies and Ten Progresses in Astronomical Science and Technology in this year.

April 8, 2019 Astronomy Department of Sun Yat-sen University became the member unit of Chinese Astronomical Society.

April 12, 2019 The Department of Science and Technology of Qinghai Province, Purple Mountain Observatory and National Astronomical Observatories signed a framework cooperation agreement on the site selection of scientific devices.

April 21, 2019 The ceremony for establishment of the Department of Astronomy at Tsinghua University was held.

June 3-5, 2019 The Seminar for National VLBI (Very Long Baseline Interferometry) Science, Technology, and Application was held at National Time Service Center.

June 5, 2019 Relying on the NAOC, co-constructed by Purple Mountain Observatory, Shanghai Observatory, and the Computer Network Information Center, the National Astronomical Data Center (NADC) had been officially included in the list of national science and technology resource sharing service platforms.

June 11, 2019 "From Compass to Beidou-Ancient Chinese Navigation Exhibition" planned by National Time Service Center was grandly opened in the Annular Hall of the United Nations Vienna International Center.

June 14, 2019 The founding meeting of the Informatizaion Working Committee, Chinese Astronomical Society (IWCC) with the information working meeting was held in Nanjing. The official website address of IWCC is unveiled: http://iwcc.china-vo.org/index.html.

June 30-July 5, 2019 The IAU Symposium 353 about galaxy dynamics in the era of large-scale sky survey was held in Shanghai.

July, 2019 A researcher team at the Timekeeping Theory and Method Research Office of National Time Service Center used self-developed coherent achromatic digital terminals and National Time Service Center's Luo Nan Haoping station's 40-meter-aperture radio telescope to obtain the fine structure of the contour of the millisecond pulsar B1937+21.

August 5-9, 2019 The international symposium "Ia Supernova and Progenitor" was held in Lijiang, Yunnan.

August 6-9, 2019 The 8th "Youth Astronomy Forum" was held in Qingdao.

August 10, 2019 Qinghai Province major science and technology special project "astronomical large scientific device cold lake site monitoring and pilot scientific research" project kick-off meeting was held in Xining City, Qinghai Province.

September 2-5, 2019 The 8th Cross-strait Symposium on Telescope and Observation Frontier Technology was held in Kaohsiung, Taiwan.

September 6-10, 2019 The 2019 annual academic meeting of Chinese Astronomical Society was held in Delinha City. More than 600 participants attended. There were 11 specially invited talks, 2 high-level popular science talks, 2 special lectures, 253 oral talks in 8 sub-meetings.

September 27, 2019 The national key basic research and development plan (973 plan) project "110-meter large-aperture fully movable radio telescope key technology research" project was successfully accepted.

September 28, 2019 The second batch of scientific results of the Dark Matter Particle Explorer satellite "Wukong" was officially released.

October, 2019 The Worldwide Telescope (WWT) Interactive Astronomy Classroom of Lingyuan No. 2 High School was completed. Lingyuan No. 2 High School was awarded the "Popular Science Education Base of The IWCC".

October, 2019 - August, 2020 The 4th WWT Guided Tour Contest was held successfully.

October 8-10, 2019 Chinese Astronomical Society carried out the selection of the 4th Huang Runqian prize, and Wang Jianming of High-Energy Physics Instirute won this prize.

October 14-21, 2019 The president of IAU, Prof. Ewine van Dishoeck of Leiton University in Netherlands, visited Yunnan Astronomical Observatory and visited Purple Mountain Observatory on October 21 2019.

October 14 - November 2, 2019 The 14th international training school of young astronomers was held in Kunming.

October 21, 2019 The second China Tianquan Lake Astronomical Forum was held in Xuyi, Jiangsu province.

October 22, 2019 The National Astronomical Observatories, Lijiang municipal government, Yunnan Observatory, and Yunnan University signed a cooperation framework agreement to build the Lijiang branch of NADC (National Astronomical Data

Center).

October 29-31, 2019 The first Symposium on Astronomy Science Education in Middle Schools of Jiangsu Province was successfully held in Nanjing University.

November, 2019 Researcher Chang Ji of PMO was elected as an academician of the Chinese Academy of Sciences, Department of Mathematical Physics.

November 4-8, 2019 The international conference on cosmology and galaxy formation was jointly organized by the Department of Astronomy at Shanghai Jiao Tong University, the Tsung-Dao Lee Institute, and Shanghai Astronomical Observatory.

November 18, 2019 Prof. Shi Shengcai of Purple Mountain Observatory, received the Ho Leung Ho Lee Foundation Award for Progress in Science and Technology.

November 18, 2019 Purple Mountain Observatory and the Key Laboratory of Planetary Science of the Chinese Academy of Sciences were awarded the "Outstanding Volunteer Unit" by the Ho Leung Ho Lee Foundation.

November 19, 2019 The Seminar of the Chinese Astronomical Society for Book Information and Periodical Publishing was held at National Time Service Center.

November 20, 2019 The domestic original 4K dome film *PRECURSOR – Guo Shoujing's astronomical achievements* was open to the public through the National Astronomical Data Center (NADC).

November 24-26, 2019 The 20th East Asia Regional Workshop on Submillimeter Receiver Technology was held in Nanjing. The meeting was sponsored by Purple Mountain Observatory and Key Laboratory of Radio Astronomy, Chinese Academy of Sciences.

November 27 - December 1, 2019 The China-VO and Astroinformatics 2019 was held successfully in Daqing, the city of lakes.

December, 2019 The 1.8 Chinese Large Solar Telescope (CLST) gets its first light in December 2019, and in debugging. This is the first 2-meter solar telescope in China.

December, 2019 Jiangsu Astronomical Society launched the selection of the first "Science and Technology Award" and "Young Talent Award".

December 7, 2019 The department of astronomy of the school of physics and astronomy of Sun Yat-sen University unveiling ceremony was held at the Zhuhai Campus of Sun Yat-sen University.

December 8-13, 2019 The Fourteenth Conference of the International Committee on Global Navigation Satellite Systems was held in Bangalore, India.

December 10, 2019 The project "Worldwide planetarium based on big data of virtual astronomical observatory" completed by members of IWCC won the nomination of the "Exhibition Award" of the 2018 science and Technology Museum Development Award.

December 16, 2019 The first National Symposium on Astronomical Photonics was held in Nanjing Institute of Astronomical Optics and Technology.

December 21, 2019 The news briefing of the result of IAU100 NameExoWorlds in the Chinese mainland was held in Beijing Planetarium.

December 26, 2019 The "Astronomy open day" was successfully held in the Worldwide Telescope interactive astronomy classroom of Lingyuan No. 2 senior high school.

December 26-27, 2019 The second symposium for female astronomers was successfully held at the Nanshan Station of XAO.

December 27, 2019 The "Maritime Silk Road Time Center" jointly constructed by National Time Service Center and the People's Government of Quanzhou City, Fujian Province was officially unveiled.

December 30, 2019 CAS carried out the selection of the 15th Zhang Yuzhe prize, and Wu Yuefang won this prize.

2020 The Beidou-3 global network was finalized, with the information processing system, the on-board hydrogen atomic clock, and the laser ranging system developed by SHAO.

2020 The year 2020 was the 60th anniversary of the founding of the astronomy discipline of Peking University, which has made important contributions to China's astronomy education and scientific research.

2020 The Xiamen University Observatory project was established.

January, 2020 The Ministry of Industry and Information Technology of the People's Republic of China officially announced the third national industrial heritage lists. The "Pucheng Long-Short Wave Radio Time Signaling Station of National Time Service Center" was successfully selected to be the only large scientific device selected by the Chinese Academy of Sciences.

January 6-10, 2020 The 5th writing school of scientific papers for young astronomers was organized by the European Journal *Astronomy and Astrophysics*, EDP Press, and Yunnan Astronomical Observatory in Kunming.

January 9, 2020 Miss Chengzhuo of Purple Mountain Observatory Named "Pawnshop - 2019 Science and Technology Jiangsu" Annual Science Communication Personality.

January 11, 2020 FAST passed the national inspection and officially opened to the public.

March 23, 2020 The near-Earth asteroid 2020 FL2, discovered by Purple Mountain Observatory, flew by Earth at 0.38 Earth-Moon distances (about 144,000 km) at 04:38:24 a.m. This is the closest near-Earth asteroid to Earth's orbit discovered by CNEOST so far.

May, 2020 Prof. Gregory J. Herczeg, Associate Director for Science for KIAA at Peking University, was named Chair of the Scientific Advisory Committee to the Thirty Meter Telescope (TMT) in 2020, and he was also appointed as the newest Science Editor for the AAS Journals.

June 21, 2020 An annular solar eclipse will be visible in the south of China, and the Astronomical Sciences Union of the Chinese Academy of Sciences (CAS) will organize solar eclipse observation and webcasting activities in many places.

June 21, 2020 An annular eclipse happened in southern China. Millions of people came to Amoy to observe.

July 25, 2020 The "Lobster-Eye X-ray Satellite" was successfully launched into orbit from the Taiyuan Launch Center, riding on the Long March 4B rocket.

July 26, 2020 Astronomy Department of Huazhong University of Science and Technology was formally established.

July 29, 2020 The Committee on Astronomical Terminology organized the first translation into Chinese of all 1,950 names of terrain features on Mars released by the International Astronomical Union (IAU), and released them freely to the public. This release includes the first batch of 814 translations of terrain names (published on 29 July), and the second batch of 1,136 translations of Martian craters (published on 24 August).

August, 2020 1m survey telescope in time domain of Nanjing University had successfully passed the inspection in Nanjing Institute of Astronomical Optics and Technology.

September 19, 2020 The national platform led by Sun Yat-sen University (CSST Science Center for the Guangdong-Hongkong-Macau Greater Bay Area) was held, providing a domestic first-class and internationally competitive scientific research platform for the cultivation of astronomical talents.

September 29, 2020 The opening ceremony of major project of science and technology "Jingdong 120m Radio Telescope for Pulsars (JRT)" was held in the site of JRT – Xujiaba, Taizhong town, Jingdong county, Pu'er City, Yunnan Province.

October 12-14, 2020 "Chinese Astronomical Society 2020 Annual Academic Conference" was held from October 12-14, 2020, for the first time online and offline. More than 1000 participants attended. There were 12 specially invited talks, 2 high-level popular science talks, 2 special lectures, 251 oral talks in 8 sub-meetings.

November, 2020 Department of Astronomy at Beijing Normal University studied and identified the magnetar origin of fast radio bursts using observations with the Five-hundred -meter Aperture Spherical radio Telescope (FAST) and the "Insight" Hard X-ray Modulation Telescope (HXMT), which was selected by the *Nature* and *Science* as one of the top ten scientific discoveries of 2020.

November, 2020 The top international journal *Metrologia* in the field of time and frequency published the academic paper "Analysis on the time transfer performance of BDS-3 signals" by the GNSS time performance evaluation team of Time and Frequency Reference Laboratory of National Time Service Center. Studies had shown that the time transfer performance of Beidou-3 has increased by more than 50% compared to Beidou-2.

November 21, 2020 The opening ceremony of the Chinese Space Station Telescope Peking University Science Center (CSST PKU Center) was held at Peking University.

November 25-29, 2020 The China-VO and Astroinformatics 2020 were held successfully in Xiamen, Ludao.

December, 2020 The SKA special cosmic dawn and reionization detection project was established.

December 6-8, 2020 The Annual Meeting and 90th Anniversary Celebration of Jiangsu Astronomical Society was held in Nanjing.

December 15, 2020 The research results of China's Sky Eye FAST telescope on fast radio bursts were selected as one of the top ten scientific discoveries of 2020 by *Nature*.

December 19, 2020 Lunar samples received by the Chang 'e-5 mission were received at the National Astronomical Observatories.

2021 *The Truth of Time* by the National Time Service Center obtained the Excellent Science Works of Minister of Science and Technology, CAS, and Anhui Province.

January, 2021 The science popularization service project of "Looking up at the Stars, Exploring new Horizons of Astronomy" of Jiangsu Astronomical Society was awarded as "Advanced Model of 2020 Science and Technology Volunteer Service Project" by China Association for Science and Technology.

January 11 - February 1, 2021 The "uncrowned king" of astronomical software, the new-year series lecture of astronomy software masters was successfully held.

January 25, 2021 Chinese Astronomical Society carried out the selection of the 8th Huang

Shoushu prize, and Tian Hui of Peking University and Yan Hongliang of NAOC won this prize.

February, 2021 According to BIPM, in 2020 the performance of China's time standard UTC (NTSC) kept by National Time Service Center will continue to be in the forefront of the world, and National Time Service Center will contribute to the International Atomic Time with a weight of 7.0%, which is the third in the world.

February 5, 2021 General Secretary Xi Jinping cordially met with the person in charge of the "China Sky Eye" project and the backbone of scientific research.

February 22, 2021 General Secretary Xi Jinping cordially met with the delegation of the engineering team of Chang'e 5 lunar exploration (director of Xinjiang Astronomical Observatory, Wang Na, representing the VLBI subsystem of orbit determination at Nanshan Station).

March, 2021 The signing ceremony of the sub center of NADC was held at the National Astronomical Observatory in Beijing. Tianjin University and Guangzhou University had signed cooperation and co-construction agreements with NADC, which will undertake the construction tasks of the NADC technology R & D and innovation center and the Greater Bay Area center of NADC, respectively.

March 31, 2021 FAST officially opened to the world.

May, 2021 Jiangsu Astronomical Society was awarded the "Chinese Social Organization Evaluation Grade 4A" by the Jiangsu Provincial Civil Affairs Department, which is the highest grade in recent developments.

May 7-9, 2021 Chinese Astronomical Society carried out the selection of the 5th Huang Runqian prize, Wang Tinggui of USTC and Zhu Zonghong of Beijing Normal University won this prize.

May 11, 2021 The groundbreaking ceremony of the University of Science and Technology of China - Purple Mountain Observatory 2.5-meter Wide Field Survey Telescope (WFST) project was held in Saishiteng Shan astronomical observatory site, Lenghu Town, Mangya City, Haixi Prefecture, Qinghai Province.

May 19, 2021 A press conference on the third batch of scientific results of the Dark Matter Particle Explorer satellite "Wukong" was held at the Purple Mountain Observatory.

May 21, 2021 The Laboratory of Microwave Technology of Xinjiang Astronomical Observatory was formally accepted as the Key Laboratory of Xinjiang Autonomous Region.

June 9, 2021 Sun Yat-sen University and National Astronomical Observatories signed the agreement to establish "United Research Center of Chinese Tianyan".

June 26, 2021 Multi-Application Survey Telescope Array project of Purple Mountain Observatory started in Cold Lake, Qinghai.

July, 2021 After years of research, the team members of the Time and Frequency Measurement and Control Laboratory of National Time Service Center had achieved a series of results in clock control, remote time comparison, and other technologies, which completed the standard time remote reproduction system engineering construction, and passed the acceptance test. It has the ability to provide domestic and foreign users with a time signal with a deviation of better than 5ns from the national standard time.

July, 2021 Shanghai planetarium, the world's largest planetarium, was officially opened. The first educational space "astronomical digital laboratory" specially reflecting astronomical digital research and data visualization application in China has been built in the museum.

August, 2021 Fujian Astronomical Society had founded Wushi Manor, an observation base in Gaiyang, Yongtai.

September 7, 2021 The National Space Science Data Center of the National Space Science Center and Purple Mountain Observatory jointly released the first batch of gamma photon scientific data from the Dark Matter Particle Explorer Satellite "Wukong".

September 25, 2021 The launching ceremony of the Lhasa timing monitoring station construction project of the high-precision ground-based timing system for a major national science and technology infrastructure project was held in the Economic

Development Zone of Lhasa City.

September 25-28, 2021 The 9th Cross-strait Astronomical Telescope and Instrument Symposium was held in Pingtang, Guizhou.

October, 2021 The 5th WWT Guided Tour Contest was officially launched, and the final award ceremony is planned in July 2022. Among the winning tours, 6 of them will be selected to participate in the first international WWT Tour Contest.

October 14, 2021 Chinese Hα Solar Explorer "Xi He" was launched on a Long March 2D carrier rocket from Taiyuan Satellite Launch Center, ushering in the era of China's space exploration of the Sun.

October 14, 2021 The 2021 National Time and Frequency Academic Conference was held in Dunhuang City, Gansu Province.

November, 2021 The "Beidou/GNSS Global Continuous Monitoring and Evaluation Data Center System Key Technology and Application" jointly applied by Wuhan University and the National Time Service Center won the first prize of the Satellite Navigation and Positioning Technology Progress Award issued by the China Satellite Navigation and Positioning Association.

November, 2021 National Natural Science Foundation of China's Major Research Instrument Development Project "2.5m Wide Field & High-resolution Telescope" had been approved.

November 18, 2021 Researcher Shi Shengcai of Purple Mountain Observatory was elected as an academician of the Chinese Academy of Sciences, Department of Mathematical Physics.

December, 2021 The China Center of IAU Office of Astronomy for Education was established in Beijing Planetarium.

December, 2021 NADC signed a cooperation agreement with Central China Normal University, Hebei Normal University and China West Normal University to jointly build an education R & D Application Center, coordinate the implementation of curriculum R & D and education and teaching practice based on real astronomical scientific data, and realize the deep integration of scientific data, information technology, and education and teaching content.

December 2-6, 2021 The 2021 annual academic meeting of Chinese Astronomical Society was held in Nanchong City, including online and offline. More than 1,000 participants attended. There were 13 specially invited talks, 2 high-level popular science talks, 2 special lectures, 379 oral talks in 9 sub-meetings.

December 10, 2021 Purple Mountain Observatory, in collaboration with Nanjing Institute of Geology and Paleontology, made an important progress in studying the first lunar samples of Chang'e-5. It shows that Chang'e-5 landing area may have several volcanic eruptions in history, which will be expected to decipher the different material compositions in the lunar mantle source region, the energy sources of volcanic magma formation and the fine spatial and temporal distribution patterns of volcanism in lunar late period.

December 16, 2021 The academic seminar on Advanced Technology of Chinese Large Aperture Optical Infrared Telescope was held in Nanjing.

2022 The largest-ever upgrade in the history of Sheshan Astronomical Observatory was completed with the joint support of Songjiang District People's Government, Science and Technology Commission of Shanghai Municipality and SHAO.

January 29, 2022 The translation of *The Complete Book of the Universe: A New Visual Guide to National Geographic* by Purple Mountain Observatory was awarded the National Outstanding Science Popularization Work of 2020.

February 10, 2022 Chinese Astronomical Society carried out the selection of the 16th Zhang Yuzhe prize, and Zhao Yongheng of National Astronomical Observatories won this prize.

February 16, 2022 It was 120th anniversary of the birth of Mr. Zhang Yuzhe, a famous astronomer and an important founder of modern Chinese astronomy, and an academician of Chinese Academy of Sciences. An event was held in Nanjing to promote the spirit of scientists and commemorate the 120th anniversary of Mr. Zhang Yuzhe's birth.

March 30, 2022 The Chinese Association for Science and Technology (CAST) named the "First Batch of National Popular Science Bases in 2021-2025", and Purple Mountain Observatory's Purple Mountain Campus, Tsingtao Observatory and Qinghai Observatory were selected at the same time.

Astronomical Achievements that Received National First, Second, and Special Prize in 2012-2022 (Including Personal Prize)

Year	Class of National Prize	Title of Item (Main Workers)
2012	Special Prize of NSTP★	Chang 'e–2 Project (NAOC★)
2012	NNS★ Prize (2nd rank)	Energy–Spectrum Excess in High–Energy Electron Cosmic Rays (Chang Jin et al.)
2013	NNS★ Prize (2nd rank)	Evolution of Large–Sample Stars and Formation of Special Stars (Han zhanwen et al.)
2014	Eleventh Prize of CYFS★	Chen Xuefei
2014	NSTP★ Prize (2nd rank)	Key Technique of Satellite–Terrestrial High–Precision Wide–Area Position Service (NAOC★)
2014	Nominees Award of National 10 Advanced Scientific Workers	Deng Xiaohua (Nanchang University)
2014	National Advanced Scientific Worker	Wu Xin (Nanchang University)
2015	National Advanced Worker	Wang Na
2016	Special Prize of NSTP★	Beidou–2 Project (NTSC★, SHAO★)
2016	NSTP★ Prize (1st rank)	Chang 'e–3 Project (NAOC★)
2017	National Medal for Innovation	Nan Rendong
2017	Honorary Title of Model of the Times	Nan Rendong
2019	Sixteenth Prize of CYFS★	Li Jing
2019	NSTP★ Prize (2nd rank)	Enhancement of Beidou Performance and Satellite–Based DM Wide–Area Enhancing Technique and Application (Cheng Junping, Cao Yueling, Gong Xiuqiang)

Year	Class of National Prize	Title of Item (Main Workers)
2020	NNS* Prize (2nd rank)	Study of Galactic Spiral–Arm Structure based on High–Precision Measurement of Maser Celestial Bodies (Xu Ye, Zheng Xingwu, Zhang Bo et al.)
2020	25th "China Youth May Fourth Medal Group".	"Wukong" satellite payload and science team (PMO*)
2021	Special Prize of NSTP*	Chang 'e–4 Project (NAOC*)
2021	Seventeenth Prize of CYTF* Team	"Team for Developing South–Pole Survey Telescope" of Yuan Xiangyan in NIAOT*

Astronomical Achievements that Received Ministerial/ Provincial First and Special Prize in 2012-2022 (Including Personal Prize)

Year	Class of Province and Minister Prize	Title of Item (Main Workers)
2011	ST* (NS*) Prize of Yunnan Province (1st rank)	Evolution of Binary Stars and Formation of Special Stars (Chen Xuefei et al., this award was actually won in 2012)
2012	NS* Prize of Jiangxi Province (1st rank)	Nonlinear Structures of Celestial Bodies (Liu Sanqiu, Wu Xin, Deng Xinfa)
2013	STP* Prize of Army (1st rank)	Beidou–2 on–orbit technique experiment (Yang Xuhai, Leihui)
2014	ST* (NS*) Prize of Yunnan Province (1st rank)	Observation and Study on Exoplanets of Symbiotic Binary and Small–Mass Celestial Bodies (Qian Shengbang et al.)
2014	STP* Prize of Shanghai City (1st rank)	VLBI Real–Time Precise Orbit Determination and Positioning on Lunar Surface of Chang 'e–3 and Soft–Landing of Jade Rabbit Lunar Rover (SHAO*)

Year	Class of Province and Minister Prize	Title of Item (Main Workers)
2015	Special ST★ Prize of Yunnan Province	Development of New Vacuum Solar Telescope and Application in Solar Observations (Liu Zhong et al.)
2015	NS★ Prize of Shanghai City (1st rank)	Exploring Formation and Evolution of Galaxies in Dark–Matter Halo (Yang Xiaohu)
2015	ST★ Prize of Guangxi Province (1st rank)	Radiation Component and Physical Origin of Gamma–ray Burst and Afterglow (Guangxi University, NAOC★)
2015	STP★ Prize of MOE★ (1st rank)	Optical and IR Telescope for Exploring Solar Eruption (Fang Cheng et al.)
2015	Special STP★ Prize of COSTIND★	Chang 'e–3 Project (NAOC★)
2015	ST★ Prize of Shanghai City (1st rank)	Exploring Formation and Evolution of Galaxies in Dark–Matter Halo (SHAO★)
2016	ST★ Prize of Guangxi Province (1st rank)	Study and Application of Key Technique for Endurance High–precision Network Cable of 500 MPa Stress Amplitude (NAOC★)
2016	6th Famous Teacher of MOE★	Xiang Shouping
2016	Education Achievement Prize of CAS★ (1st rank)	Exploration and practice for talent class combined with science and education and training mode of physical innovation talents (Astronomy Department of USTC)
2017	ST★ Prize of Beijing City (1st rank)	Innovation and Practice of Super–Large Spatial Structure for 500 M Spherical Surface Radio Telescope Project (Nan Rendong et al.)
2017	Outstanding ST★ Achievement Prize of CAS★	Research Team of 500 M Spherical Surface Radio Telescope (FAST) Project (Nan Rendong et al.)
2017	STP★ Prize of Xinjiang Uygur Autonomous Region (1st rank)	Observation and Study on Instability of Pulsar Rotation (XAO★)
2017	Excellent NS★ Achievement Prize of MOE★ (1st rank)	Black Hole with Maximum Mass in the Brightest Quasar of Early Universe (Wu Xuebing et al.)
2017	STP★ Prize of Army (1st rank)	Key Technique and Application of Daytime High–Resolution Photo–Electronic Imaging for Space Objects (NAOC★)
2018	Special STP★ Prize of Shanghai City	Development of Shanghai 65 M Radio Telescope System (SHAO★)
2018	STP★ Prize of Liaoning Province (1st rank)	Technique and Equipment of Flexible Parallel Cable Driving System for 500 M Radio Telescope (Zhu Wenbai et al.)
2018	ST★ Prize of Jiangsu Province (1st rank)	"Wukong" Detector of Dark–Matter Particles (Chang Jin et al.)
2019	ST★ Prize of Jiangsu Province (1st rank)	Core Innovation and Key Technique of LAMOST (Su Dingqiang et al.)

Year	Class of Province and Minister Prize	Title of Item (Main Workers)
2019	Outstanding ST* Achievement Prize of CAS*	Research Team of LAMOST Project (Su Dingqiang et al.)
2019	STP* Prize of Guizhou Province (1st rank)	Application Study and Promotion of Key Technique for Excavation System of FAST (Zhu Boqin et al.)
2019	STP* Prize of Anhui Province (1st rank)	Positioning Device of Optical Fiber on Focal Plane of LAMOST (Zhao Yongheng et al.)
2019	Special Education Achievement Prize of Anhui Province	Exploration of new mode for training astronomical talents combined of science and education (Yuan Yefei et al.)
2019	STP* Prize of Guizhou Province (1st rank)	Technical Innovation and Practice of Active Reflector of FAST (Jiang Peng et al.)
2019	Special ST* (NS*) Prize of Yunnan Province	Electromagnetic Interaction in Solar Eruptions (Lin Jun et al.)
2019	Special STP* Prize of COSTIND*	Chang 'e–4 Project (Li Chunlai et al.)
2019	ST* Prize of Shanghai City (1st rank)	Passive–Mode Hydrogen Atomic Clock (SHAO*)
2019	ST* Prize of Army (1st rank)	Atomic Frequency Standard of Small Light–Pumping Cesium Beam (Zhang Shougang et al.)
2019	Xinjiang Youth May Fourth Medal	Engineering Team of Lunar Exploration of Xinjiang Astronomical Observatory
2020	STP* Prize of Hebei Province (1st rank)	Key Technique and Application of Hydraulic Element/System and Growing Reliability of Systematic Group (Zhu Ming et al.)
2020	Famous Teacher in "10000 Strong–Program" of MOE*	Li Xiangdong
2021	Outstanding ST* Achievement Prize of CAS*	Research Team of Ground–Based Application System for Lunar and Deep Space Exploration (Li Chunlai et al.)
2021	Famous Teacher and Team of Ideological and Political Education in the Curriculum of MOE*	Li Xiangdong et al.
2021	Special Educational Achievement Prize of Jiangsu Province	Exploration and Practice of Ideological and Political Education from "Exploration of Astronomical Secrets" to "Brief History of Universe" (Li Xiangdong et al.)
2021	ST* (NS*) Prize of Yunnan Province (1st rank)	Fine Physical Processes of Transportation and Release of Solar Active Energies (Yan Xiaoli et al.)
2021	Advanced Worker of CAS*	Zhang Shougang
2021	The Most Beautiful Scientist of Jiangsu Province	Yang Ji
2021	Excellent Pop–Science Book of CAS	"The Truth of Time" (Li Xiaohui)

***Abbreviations**

ST: Science and Technology

STP: Science and Technology Progress

NSTP: National Science and Technology Progress

NS: Nature Science

NNS: National Nature Science

MOE: Minister of Education

COSTIND: Commission of Science, Technology, and Industry for National Defense

NAOC: National Astronomical Observatories of China

CAS: Chinese Academy of Sciences

PMO: Purple Mountain Observatory

SHAO: Shanghai Astronomical Observatory

YAO: Yunnan Astronomical Observatory

NTSC: National Time Service Center

XAO: Xinjiang Astronomical Observatory

NIAOT: Nanjing Institute of Astronomical Optics and Technology

CYFS: Chinese Young Female Scientist

图片部分
PICTURE

一. 学会活动 Activities of Chinese Astronomical Society

2011 年 10 月，中国天文学会天文学名词审定委员会在江西南昌召开第九届第一次会议。

The Ninth Astronomical Terminology Committee of the Chinese Astronomical Society was formally established and held their first meeting in Nanchang, Jiangxi in 2011.10.

天文学名词审定委员会第九届第三次会议于 2013 年 10 月 25 日至 27 日在苏州召开。

The third meeting of the ninth session of the Astronomical Terminology Committee was held in Suzhou from 25 to 27 October 2013.

2014 年 10 月 27 日，中国天文学会 2014 年学术年会暨第十三次全国会员代表大会在陕西省西安市临潼区隆重开幕。

On October 27, 2014, Academic Annual Meeting of Chinese Astronomical Society and the 13th National Member Representative Conference were grandly opened in Lintong District, Xi'an, Shaanxi Province.

中国天文学会天文学名词审定委员会第十届委员会、全国科学技术名词审定委员会天文学名词审定委员会第七届委员会成立，第一次会议于 2015 年 10 月 21 日至 23 日在国家天文台兴隆观测站召开。

The tenth session of the Astronomical Terminology Committee of the Chinese Astronomical Society was established, and the first meeting was held at the Xinglong Observatory of the National Astronomical Observatory on 2015.10.21.

中国天文学会 2017 年学术年会（2017 年 8 月，乌鲁木齐）。

2017 annual academic meeting of Chinese Astronomical Society (August, Urumqi).

2017 年 11 月 16 日，天文学名词审定委员会第十届第二次会议在南京大学召开。

The second meeting of the 10th session of the Astronomical Terminology Committee was held at Nanjing University, Nanjing, Jiangsu on 2017.11.16.

中国天文学会第十四届（2017–2018 年度）张钰哲奖评审会议（2018 年 9 月，盱眙）。

2018 review meeting of the Chinese Astronomical Society 14th Zhang Yuzhe prize (September, Xuyi).

中国天文学会 2018 年学术年会（2018 年 10 月，昆明）。

2018 annual academic meeting of Chinese Astronomical Society (October, Kunming).

2018 年 10 月 27 日，首届中国天文学会女天文工作者交流会议部分代表合影。

The first symposium for female astronomers was held on 27 October 2018.

2019 年 6 月 14 日，信息委成立大会暨信息化工作交流研讨会在南京召开。

On June 14, 2019, the founding meeting of IWCC (Informatizaion Working Committee, Chinese Astronomical Society) and the information work exchange seminar were held in Nanjing.

中国天文学会 2019 年学术年会（2019 年 9 月，德令哈）。

2019 annual academic meeting of Chinese Astronomical Society (September, Delingha).

2019 年 11 月 19 日，中国天文学会 2019 图书信息与期刊出版研讨会在国家授时中心召开。

On November 19, 2019, the Chinese Astronomical Society 2019 Book Information and Periodical Publishing Seminar was held at National Time Service Center.

中国天文学会信息化工作委员会第二次工作会议
2020.11.24 厦门

2020 年 11 月 24 日，信息委第一届委员会第二次工作会议参会委员合照。

On November 24, 2020, the group photo of the second work meeting of IWCC.

全国科学技术名词审定委员会、中国天文学会
第十一届天文学名词审定委员会第二次会议
2020.11.28-12.1 厦门

2020 年 11 月 28 日，天文学名词审定委员会第十一届第二次会议在福建厦门召开。

The second meeting of the eleventh session of the Astronomical Terminology Committee was held in Xiamen, Fujian on 2020.11.28.

中国天文学会信息化工作委员会2021年工作会议
2021.6 江苏昆山

2021 年 6 月，信息委第一届委员会第三次工作会议在昆山召开。

In June 2021, the group photo of the third work meeting of IWCC.

中国天文学会2021年学术年会合影
西华师范大学 2021.12

中国天文学会 2021 年学术年会（2021 年 10 月，南充）。

2021 annual academic meeting of Chinese Astronomical Society (October, Nanchong).

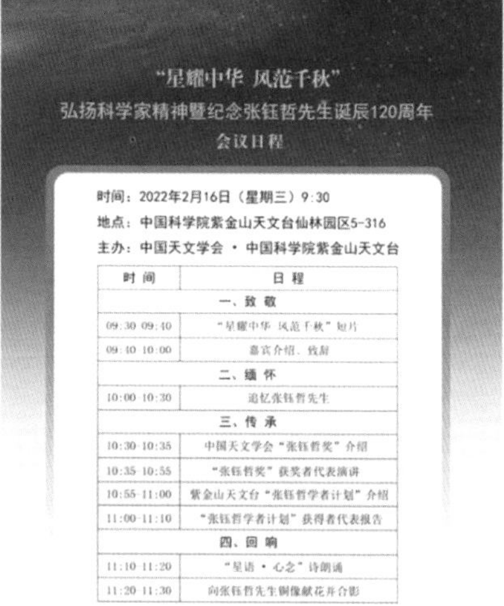

“星耀中华 风范千秋”

弘扬科学家精神暨纪念张钰哲先生诞辰120周年

会议日程

时间：2022年2月16日（星期三）9:30

地点：中国科学院紫金山天文台仙林园区5-316

主办：中国天文学会 · 中国科学院紫金山天文台

时 间	日 程
一、致 敬	
09:30-09:40	“星耀中华 风范千秋”短片
09:40-10:00	嘉宾介绍、致辞
二、缅 怀	
10:00-10:30	追忆张钰哲先生
三、传 承	
10:30-10:35	中国天文学会“张钰哲奖”介绍
10:35-10:55	“张钰哲奖”获奖者代表演讲
10:55-11:00	紫金山天文台“张钰哲学者计划”介绍
11:00-11:10	“张钰哲学者计划”获得者代表报告
四、回 响	
11:10-11:20	“星语 · 心念”诗朗诵
11:20-11:30	向张钰哲先生铜像献花并合影

2022 年 2 月 16 日，“星耀中华 风范千秋”弘扬科学家精神暨纪念张钰哲先生诞辰 120 周年活动在南京举办。

On 2022.2.16 an event was held in Nanjing to promote the spirit of scientists and commemorate the 120th anniversary of Mr. Zhang Yuzhe’s birth.

二. 院士风采 Academicians

叶叔华院士

Academician Ye Shuhua

曲钦岳院士

Academician Qu Qinyue

陈建生院士

Academician Chen Jiansheng

苏定强院士

Academician Su Dingqiang

熊大国院士

Academician Xiong Darun

欧阳自远院士

Academician Ouyang Ziyuan

艾国祥院士

Academician Ai Guoxiang

方成院士

Academician Fang Cheng

朱能鸿院士

Academician Zhu Nenghong

孙义燧院士

Academician Sun Yisui

李惕碚院士

Academician Li Tibei

潘君骅院士

Academician Pan Junhua

崔向群院士

Academician Cui Xiangqun

武向平院士

Academician Wu Xiangping

汪景琇院士

Academician Wang Jingxiu

景益鹏院士

Academician Jing Yipeng

韩占文院士

Academician Han Zhanwen

常进院士

Academician Chang Jin

史生才院士

Academician Shi Shengcai

三、重大项目与成果 Major Achievements and Projects

2012 年 7 月，国家天文台与贵州省科技厅共建贵州射电天文台协议签署。
The cooperation agreement between NAOC and Dept of Science and Technology of Guizhou Province to construct jointly Guizhou Radio Astronomical Observatory was signed in 2012.07.

2012 年 10 月 28 日，上海 65 米射电望远镜落成仪式及上海天文台成立 50 周年暨建站 140 周年庆典活动在松江佘山上海 65 米射电望远镜现场隆重举行。
The inauguration ceremony of the Shanghai 65m Radio Telescope and the celebration of the 50th anniversary of the establishment of the Shanghai Astronomical Observatory of the Chinese Academy of Sciences and the 140th anniversary of the establishment of the station were held in Sheshan, Songjiang on 2012.10.28.

2012年11月19日 紫金山天文台与荷兰代尔夫特理工大学（TU Delft）科学与技术合作谅解备忘录签约仪式在南京举行。

The signing ceremony of the Memorandum of Understanding on Science and Technology Cooperation between the Purple Mountain Observatory and the Delft University of Technology (TU Delft) in the Netherlands was held in Nanjing on 2012.11.19.

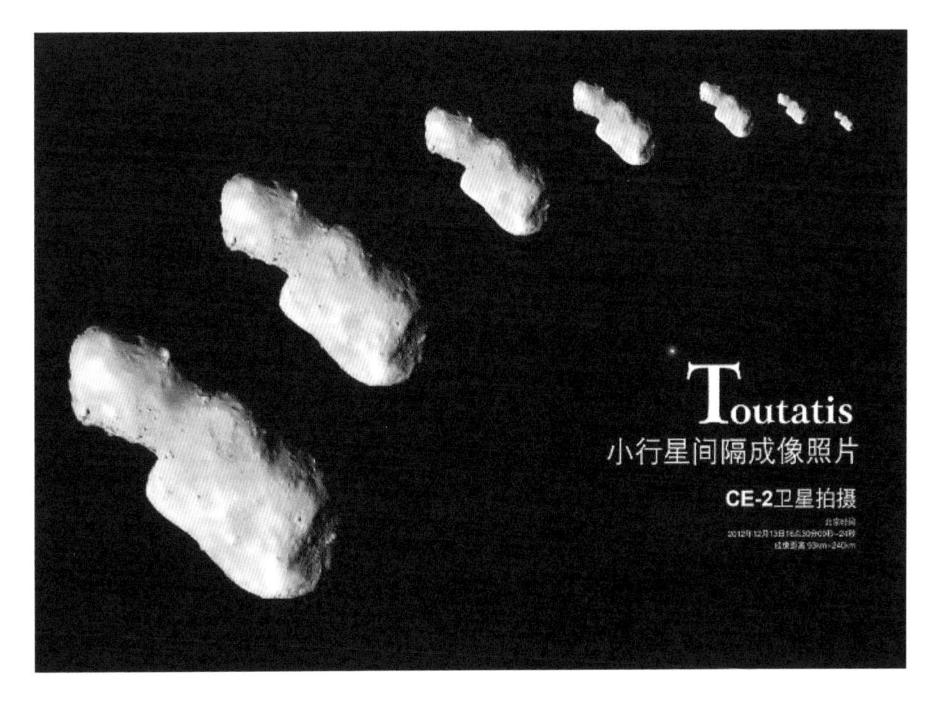

2012年12月13日，“嫦娥二号”卫星成功飞越小行星“战神”图塔蒂斯，在国际上第一次拍下这颗小行星的光学图像。紫台自主确定了图塔蒂斯轨道，圆满地完成“嫦娥二号”再拓展试验任务。

On 2012.12.13, Chang’e-2 satellite successfully flew over the asteroid “God of War” Toutatis, and took the images of this asteroid for the first time in the international arena. PMO independently determined the Toutatis orbit and successfully completed the “Chang’e-2” re-expansion test mission.

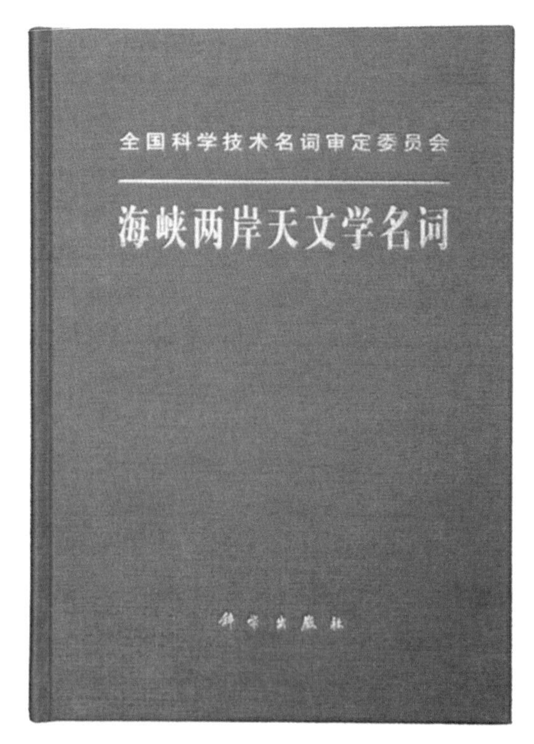

2013 年 10 月，《海峡两岸天文学名词》由科学出版社正式出版发行。

the Cross–Strait Astronomical Terminology was officially published by Science Press in 2013.10.

2014 年 ONSET 望远镜（右）成功验收，入选 2014 年度“十大天文科技进展”。

In 2014 Onset (rightside) has been accepted and selected as one of “Ten Progresses of Astronomical Sciences and Technologies”.

2014 年南极第二台巡天望远镜 AST3–2 安装成功。

The second survey telescope in South Pole, AST3-2, was successfully mounted in 2014.

国家授时中心 2014 年建成的卫星导航系统评估系统。
The construction of the evaluation system of the satellite navigation system was completed by National Time Service Center in 2014.

国家授时中心 2014 年建成的卫星导航综合试验系统。
The construction of the satellite navigation integrated test system was completed by National Time Service Center in 2014.

2015 年 1 月南极 DOME A 上的天文设备。

The astronomical devices on DOME A of South Pole in 2015.01.

2015 年 12 月 1 日，新版《英汉天文学名词》由中国科协三峡科技出版资助中国科学技术出版社出版。

The new edition of *English–Chinese Astronomical Terms* was published by China Science and Technology Press with the support of the CCSA Sanxia Science and Technology Publication Grant Scheme on 2015.12.1.

2015 年 12 月 17 日，暗物质粒子探测卫星“悟空”号在酒泉卫星发射中心用“长征二号”丁运载火箭成功发射。

Dark Matter Particle Explorer Satellite “Wukong” was successfully launched by the Long March 2D vehicle from the Jiuquan Satellite Launch Center on 2015.12.17.

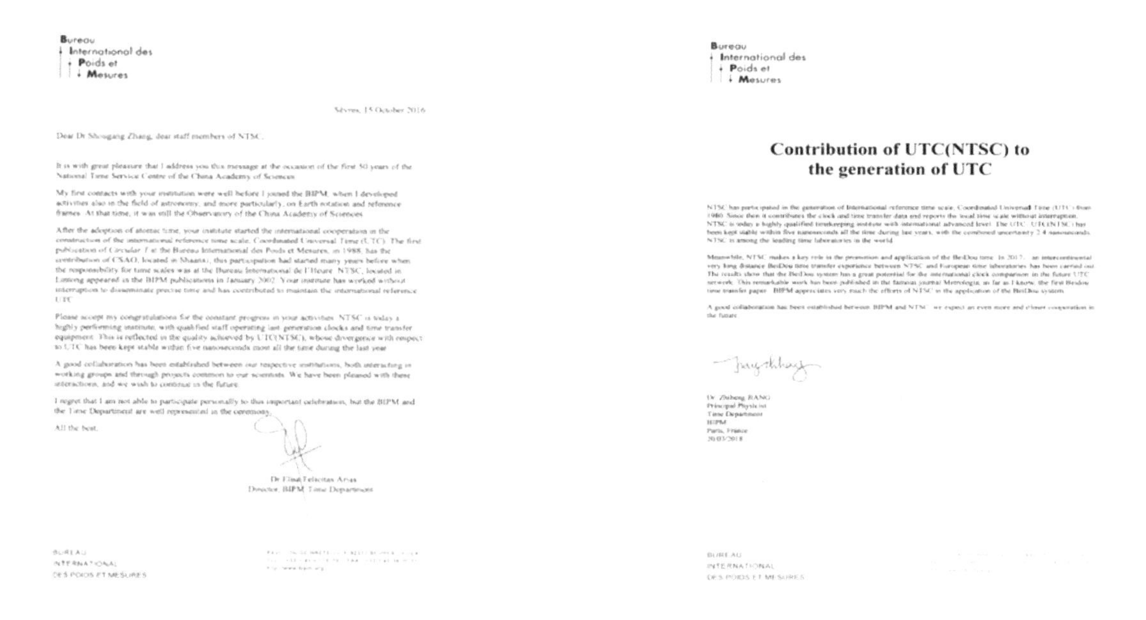

Bureau International des Poids et Mesures

Sèvres, 15 October 2016

Dear Dr Shougang Zhang, dear staff members of NTSC,

Dr Elisa Felicitas Arias
Director, BIPM Time Department

BUREAU INTERNATIONAL DES POIDS ET MESURES

Bureau International des Poids et Mesures

Contribution of UTC(NTSC) to the generation of UTC

Dr. Zhiheng JIANG
Principal Physicist

BUREAU INTERNATIONAL DES POIDS ET MESURES

左图：2016 年国际权度局（BIPM）发来的贺信，祝贺国家授时中心成立 50 周年。

右图：2018 年国际权度局 (BIPM) 来函高度赞扬国家授时中心守时水平并且充分肯定国家授时中心北斗国际时间比对工作。

Left: A congratulatory letter from BIPM in 2016, congratulating the 50th anniversary of the establishment of National Time Service Center.

Right: A letter from BIPM in 2018 which highly praised the time-keeping performance of National Time Service Center and fully affirmed the Beidou International Time Comparison Work of National Time Service Center.

2016 年 6 月 1 日，国家“十二五”科技创新成就展在北京展览馆举行，国家授时中心参展。展板展示“我国时间基准保持实现国际先进水平”。

On June 1, 2016, the National “12th Five-year” Science and Technology Innovation Achievement Exhibition was held in Beijing Exhibition Hall. National Time Service Center participated in the exhibition. This board showed that “our national time standard has maintained an internationally advanced level.”

2016 年 8 月 19 日，云南天文台 – 香港天文学会宽视场测光巡天项目正式开始运行。

The wide-field photometric survey project of Yunnan Astronomical Observatory and Hong Kong Astronomical Society started to operate on 2016.8.19.

2016 年 9 月 25 日，500 米口径球面射电望远镜（FAST）工程竣工。

On September 25, 2016, the 500-meter Aperture Spherical Radio Telescope (FAST) project was completed.

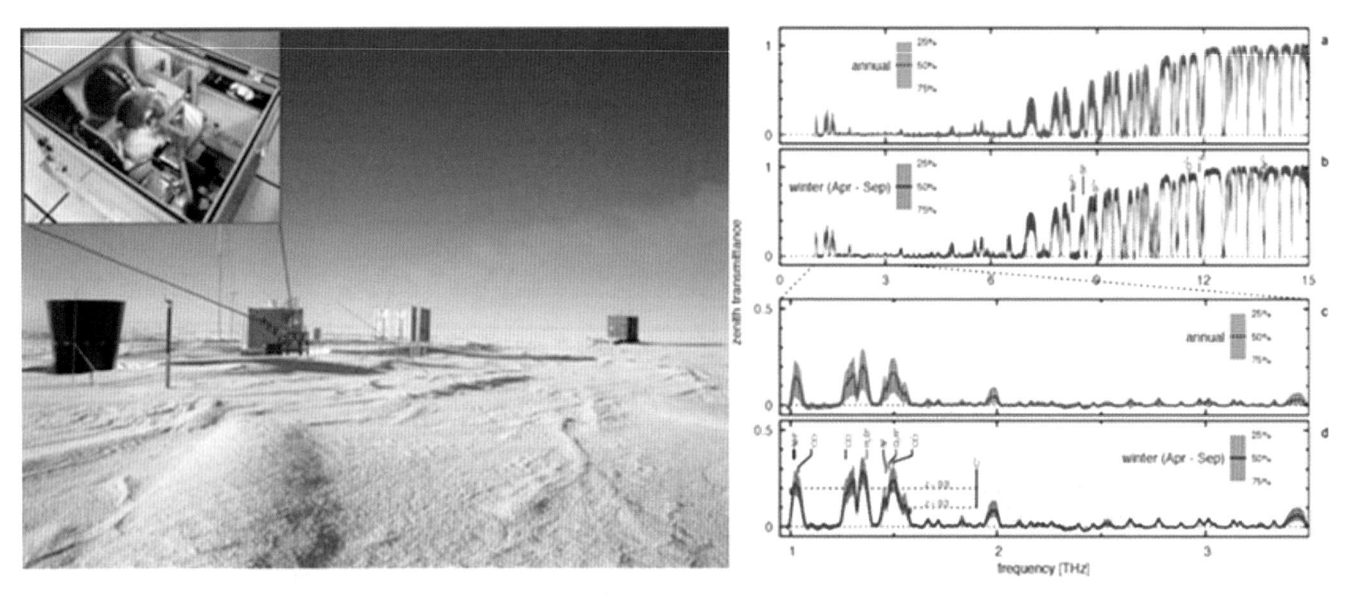

2016 年 12 月 13 日，国际上首例以无人值守工作模式运行的超宽带（0.75~15THz）太赫兹傅里叶光谱仪获得了南极冰穹 A 天文台址太赫兹至远红外谱段的大气透过率长周期实测数据。

The world's first ultra-wideband (0.75~15THz) terahertz Fourier spectrometer operating in unattended working mode was successfully installed and obtained the long-period measured data of atmospheric transmittance in the terahertz to far-infrared spectrum at the Dome A Observatory in Antarctica on 2016.12.13.

2017 年 10 月，被誉为下一代大视场自适应光学的多层共轭自适应光学试验系统在云南天文台 1 米新真空太阳望远镜（NVST）上成功闭环，在国内首次实时获得太阳活动区大视场高分辨力图像。

In October 2017, a solar Multi-Conjugate Adaptive Optics (MCAO) experiment system was, for the first time, built and successfully obtained the first light based on 1-meter New Vacuum Solar Telescope in China.

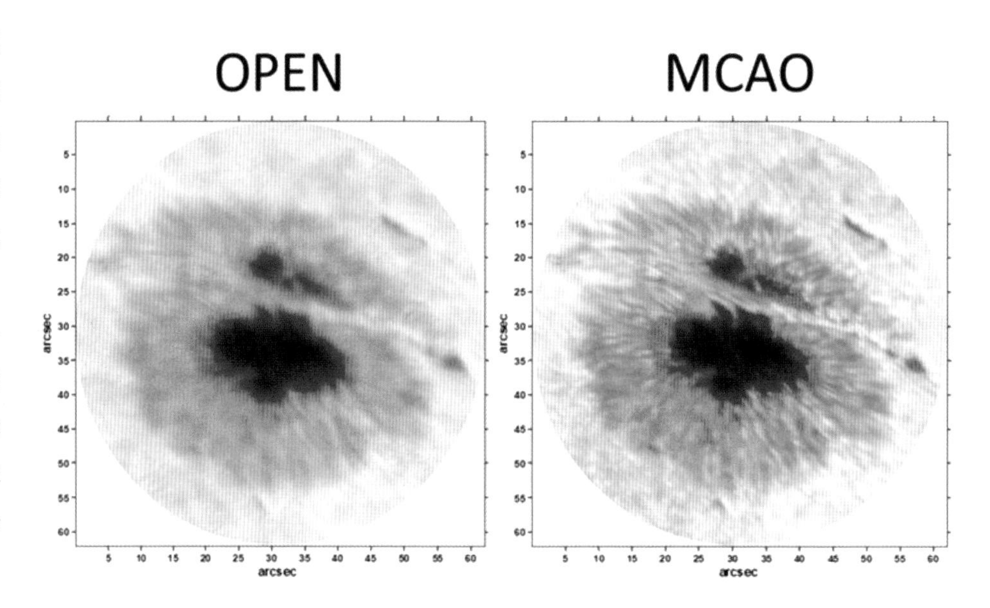

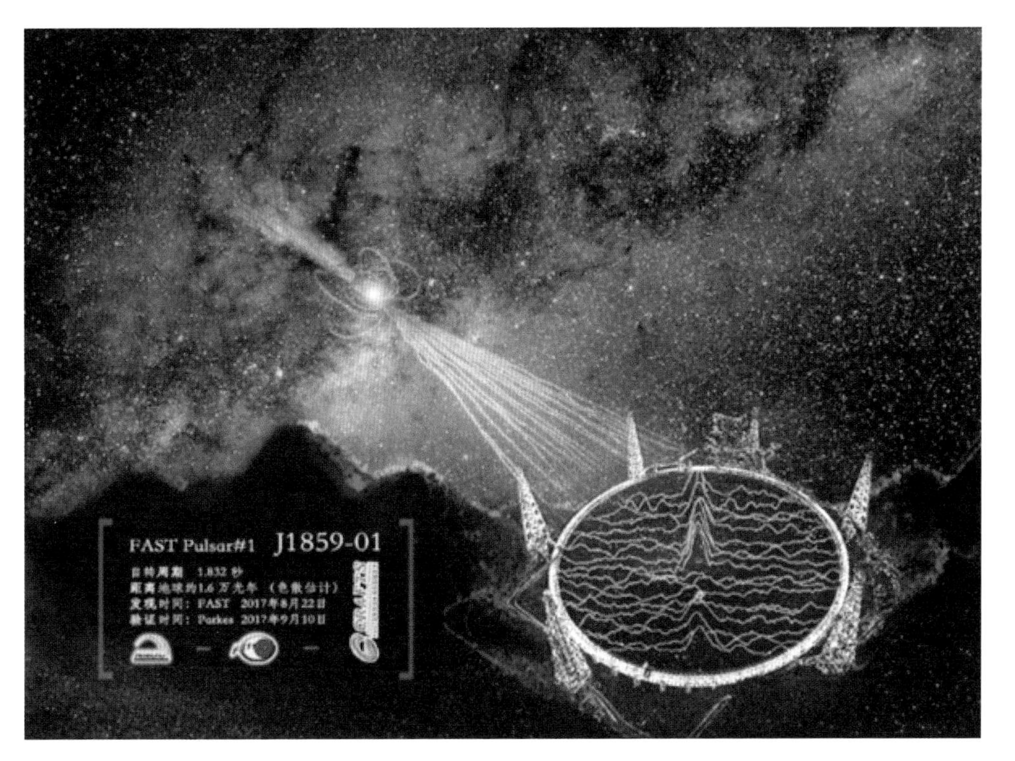

2017 年 10 月 10 日，由中国科学院主持发布 500 米口径球面射电望远镜（FAST）取得的首批成果，包括 6 颗新脉冲星。

The first results of the 500-meter Aperture Spherical Radio Telescope (FAST), including 6 new pulsars, were released under the auspices of the Chinese Academy of Sciences on 2017.10.10.

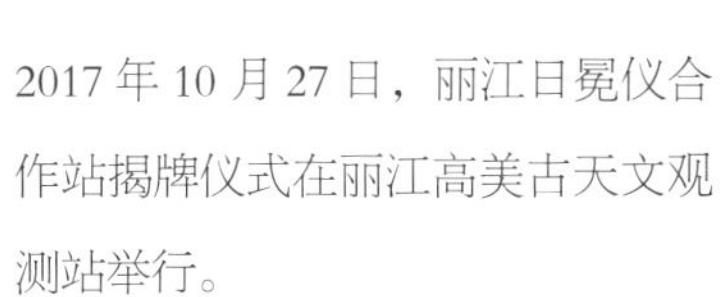

2017 年 10 月 27 日，丽江日冕仪合作站揭牌仪式在丽江高美古天文观测站举行。

The cooperated station of coronagraph was opened in Lijiang, Gaomeigu station
on 2017.10.27.

2017 年 11 月 16 日，新版英汉天文学词典正式上线，纳入中国虚拟天文台平台。

A new edition of the English-Chinese Astronomy Dictionary was officially launched and incorporated into the China Virtual Observatory platform on 2017.11.16.

2018 年，国产小型光抽运小铯钟获得了中国国际工业博览会技术创新金奖。

2019 年，获得国际权度局认可，向全世界推荐。

In 2018, the domestically manufactured small optically pumped cesium beam atomic clock won the Technology Innovation Gold Award of China International Industry Fair.

In 2019, the domestically manufactured small optically pumped cesium beam atomic clock was recognized by BIPM and was recommended to the world.

2019 年 9 月 23 日，“伟大历程 辉煌成就——庆祝中华人民共和国成立 70 周年大型成就展”在北京展览馆开幕。国家授时中心策划的“国家授时体系建设”亮相本次展览“1981 年展区”。

On September 23, 2019, “The Construction of National Time Service System” planned by National Time Service Center was exhibited in the “Great History and Brilliant Achievement–Large Achievement Exhibition to Celebrate the 70th Anniversary of the Founding of the People's Republic of China”.

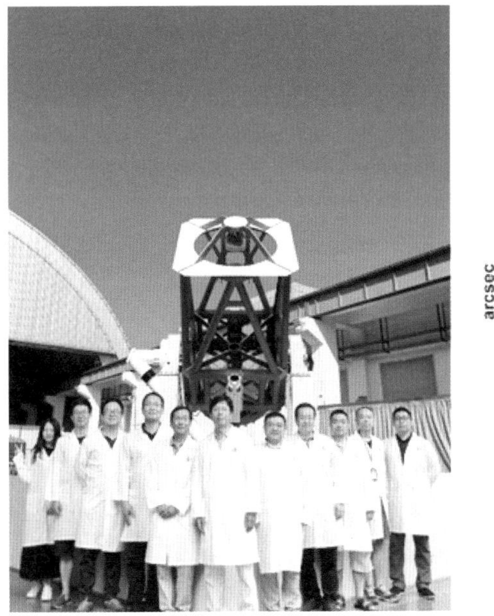

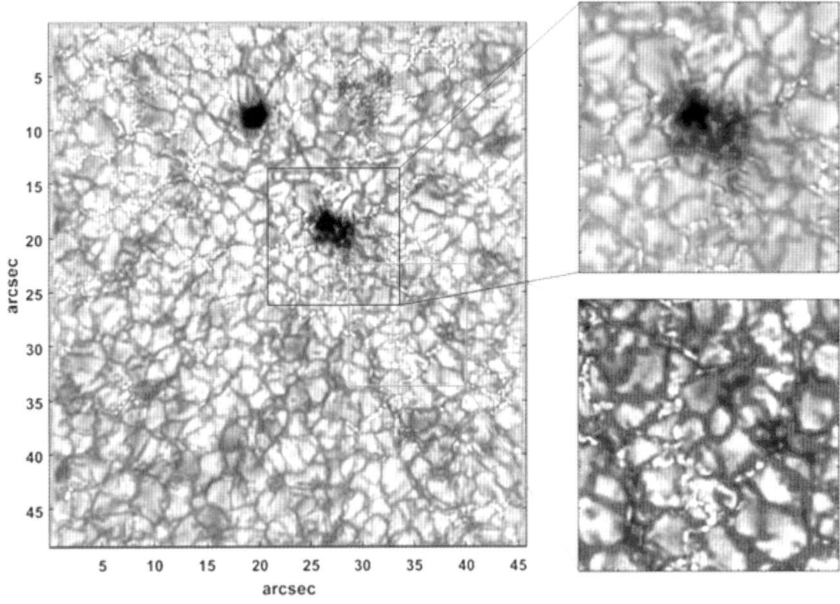

2019 年 12 月，1.8 米太阳望远镜（Chinese Large Solar Telescope, CLST）实现初光并在调试中。该望远镜是国内首台 2 米级口径太阳望远镜。

The 1.8 Chinese Large Solar Telescope (CLST) got its first light in December 2019, and in debugging. This is the first 2-meter solar telescope in China.

2020 年南京天光所承担的中国科学院“一带一路”科技合作专项项目验收。
In 2020 the specific project of CAS science and technology cooperation for “Belt and Road Initiative” by Nanjing Institute of Astronomical Optics and Technology had been accepted.

2020 年 1 月 11 日 FAST 通过国家验收，正式开放运行。
On January 11 2020 FAST passed the national inspection and officially opened to the public.

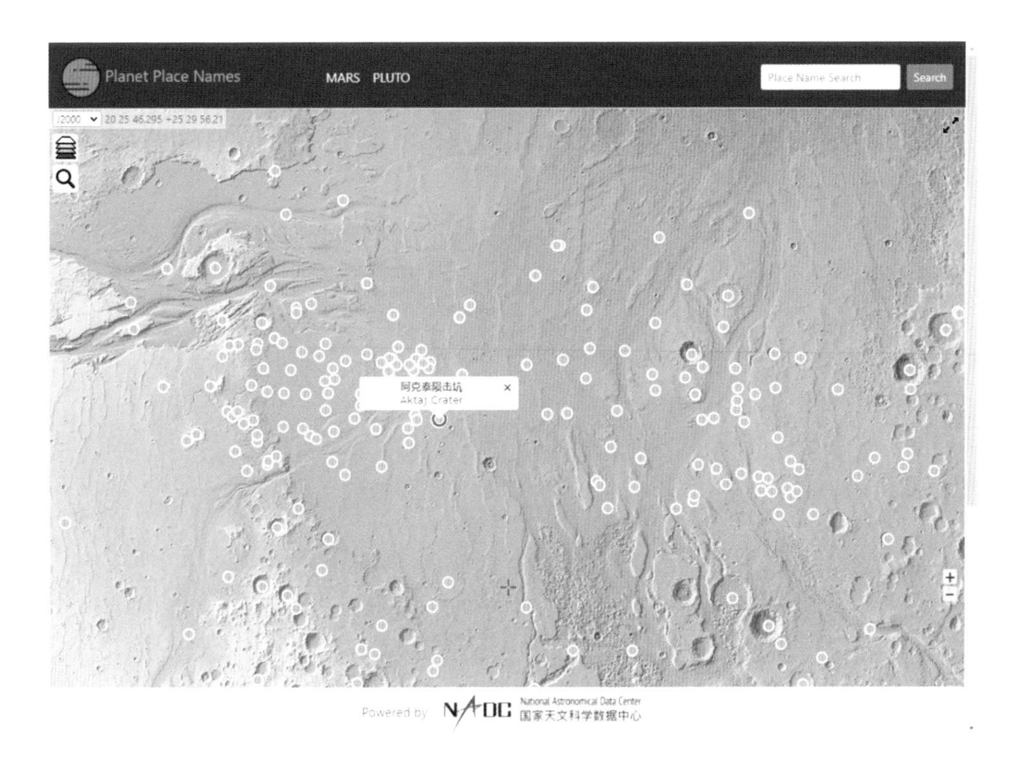

2020 年 7 月 29 日，天文学名词委组织力量首次将国际天文学联合会公布的全部 1950 条火星地形地貌名称全部译为中文，并向全社会公开发布。

The Committee on Astronomical Terminology organised the first translation into Chinese of all 1950 names of terrain features on Mars released by the International Astronomical Union (IAU), and released them freely to the public on 2020.7.29.

2020 年 9 月 19 日，以中山大学为牵头单位的国家级平台——中国空间站工程巡天望远镜（CSST）粤港澳大湾区科学中心在学院举行了揭牌仪式。

The national platform led by Sun Yat-sen university(CSST Science Center for the Guangdong-Hongkong-Macau Greater Bay Area) was held on September 19, 2020.

2020 年 11 月 21 日，中国空间站工程巡天望远镜北京大学科学中心揭牌成立。

On November 21, 2020, the opening ceremony of the Chinese Space Station Telescope Peking University Science Center (CSST PKU Center) was held at Peking University.

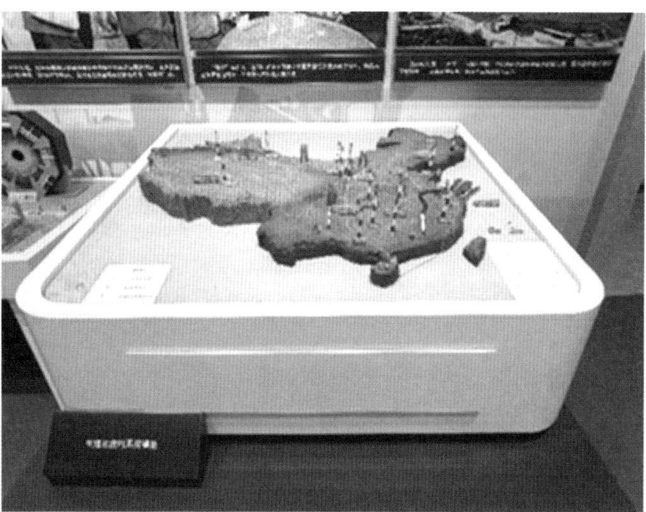

长短波授时系统亮相 2021 年开馆的中国共产党历史展览馆。

The long-wave and short-wave radio time signal system is exhibited in CPC Museum which was open in 2021.

2021 年，“司天工程”Mini 阵望远镜项目完成台站验收。

In 2021, the Mini Array of Telescopes for the project of “Celestial Manager” had been accepted.

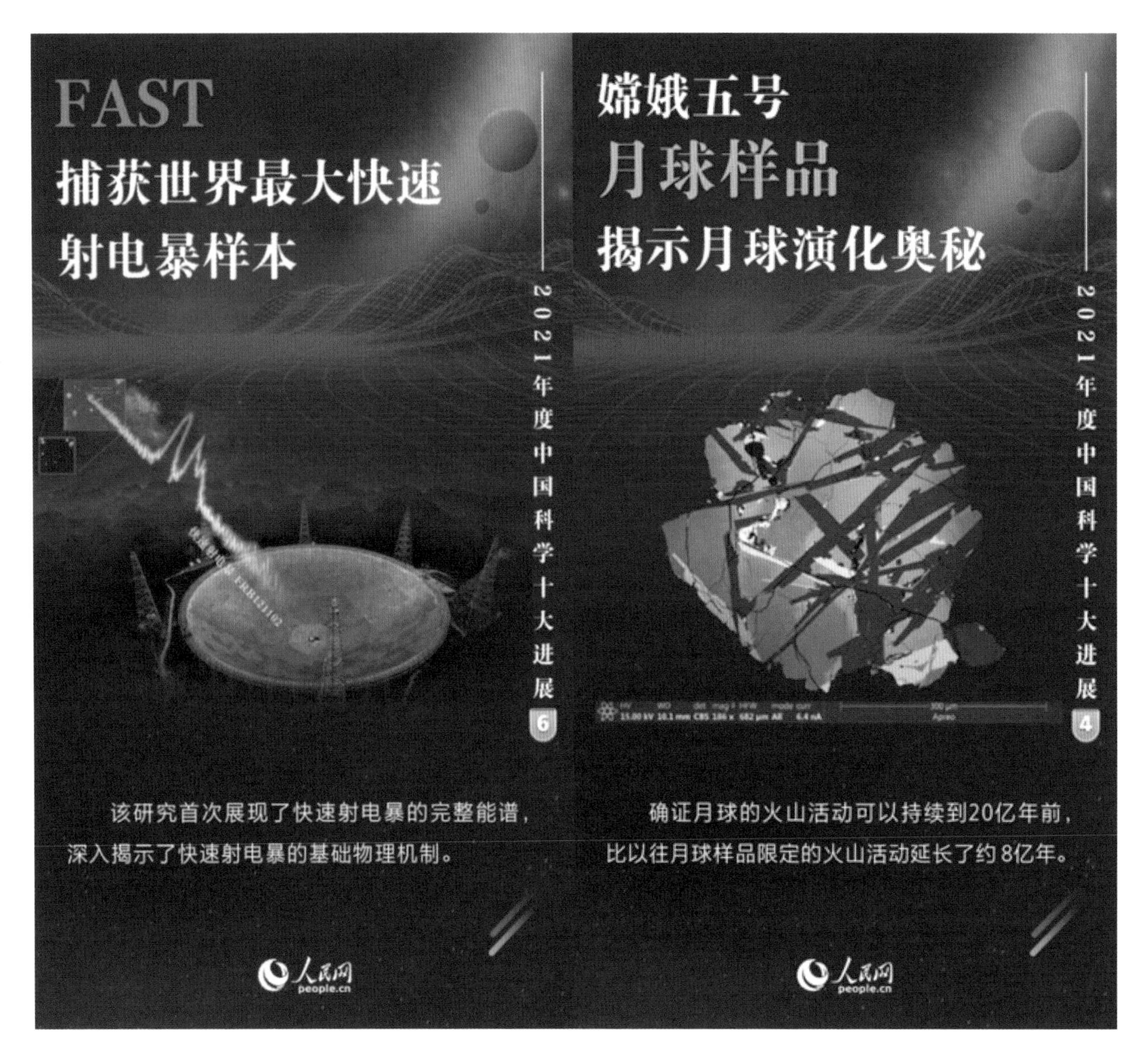

2021 年，FAST 捕获最大快速射电暴、“嫦娥五号”月球样品揭示的月球演化奥秘分别评为中国科学十大进展之一。

In 2021 the largest fast radio burst catched by FAST and the mistery of lunar evolution explored by the lunar sample collected by Chang’e-5 had been evaluated respectively as one of 10 major progresses of Chinese Sciences.

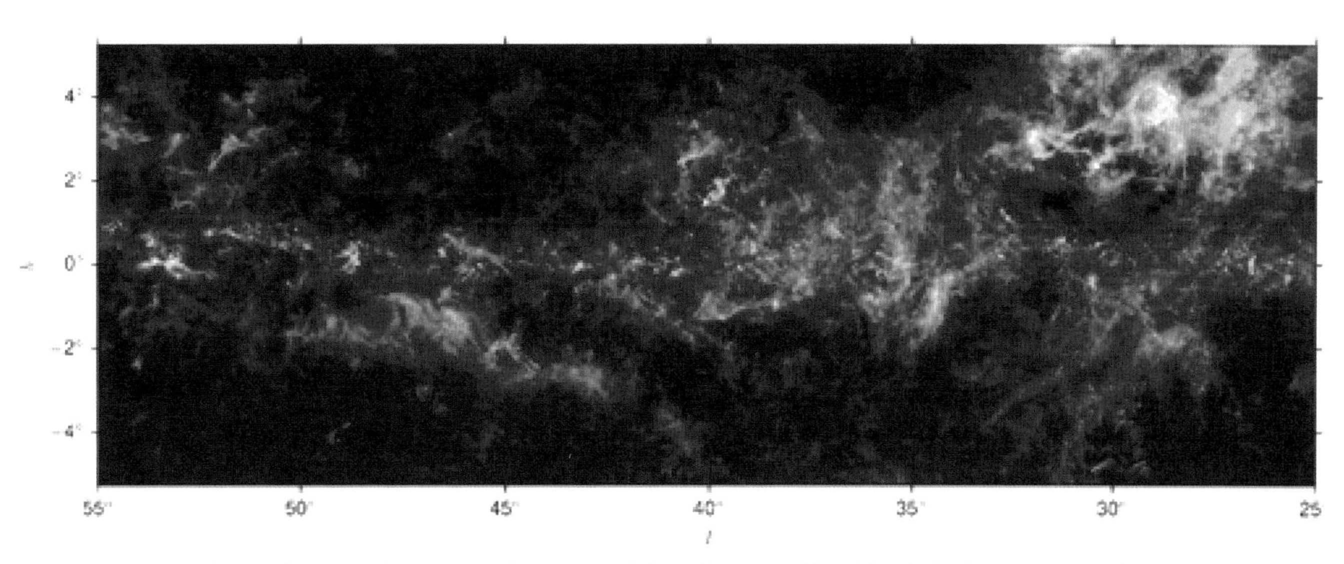

2021 年 4 月，紫金山天文台从 2011 年 11 月开始组织开展“银河画卷”巡天计划，使用青海观测站 13.7 米的毫米波望远镜，对北天银道面进行大天区观测，一期巡天历时 10 年。

Purple Mountain Observatory started to organize the “Milky Way Picture Scroll” survey program in November 2011, using a 13.7-meter millimeter telescope at the Qinghai Observatory to observe the northern celestial silver channel surface in a large sky area.

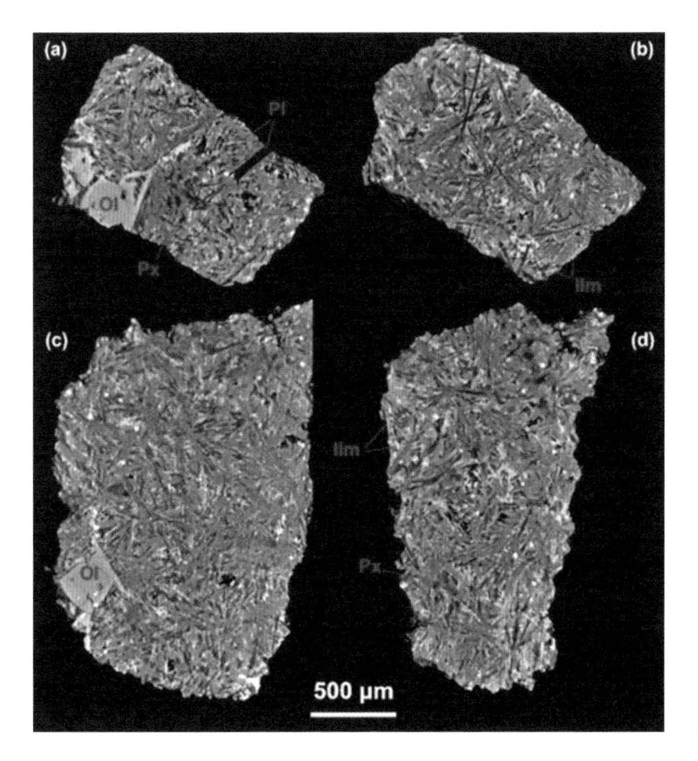

2021 年 12 月 10 日，紫金山天文台联合南京地质古生物研究所合作研究“嫦娥五号”首批月球样品取得重要进展。

Purple Mountain Observatory, in collaboration with Nanjing Institute of Geology and Paleontology on 2021.12.10, made an important progress in studying the first lunar samples of Chang’e−5. It shows that Chang’e−5 landing area may have several volcanic eruptions in history.

四、重要天象的观测 Observations of Important Celestial Events

2012 年 ONSET 拍摄的 3600 埃全日面白光太阳像“金星凌日”。

In 2012, the transit of Venus in the solar full-disk white-light image by ONSET at 3600 Å.

2012 年 5 月 21 日，我国东南沿海出现日环食，国内外多支队伍莅闽，图为中日两国天文爱好者在海上丝路起点——福建石狮市石湖蚶江港共同观测。

On 21 May 2012, an annular eclipse happened upon the southern and eastern sea of China. Numbers of astrophile teams came to Fujian. The picture shows the scene of astrophiles from China and Japan carrying out observational work in Hanjiang Harbor.

2017 年 8 月 21 日，云南天文台组织的观测组在美国俄勒冈州达拉斯市对日全食进行了成功观测。

The group organized by Yunnan Astronomical Observatory successfully observed the full eclipse in Dallas of Oregon State of USA on 2017.8.21.

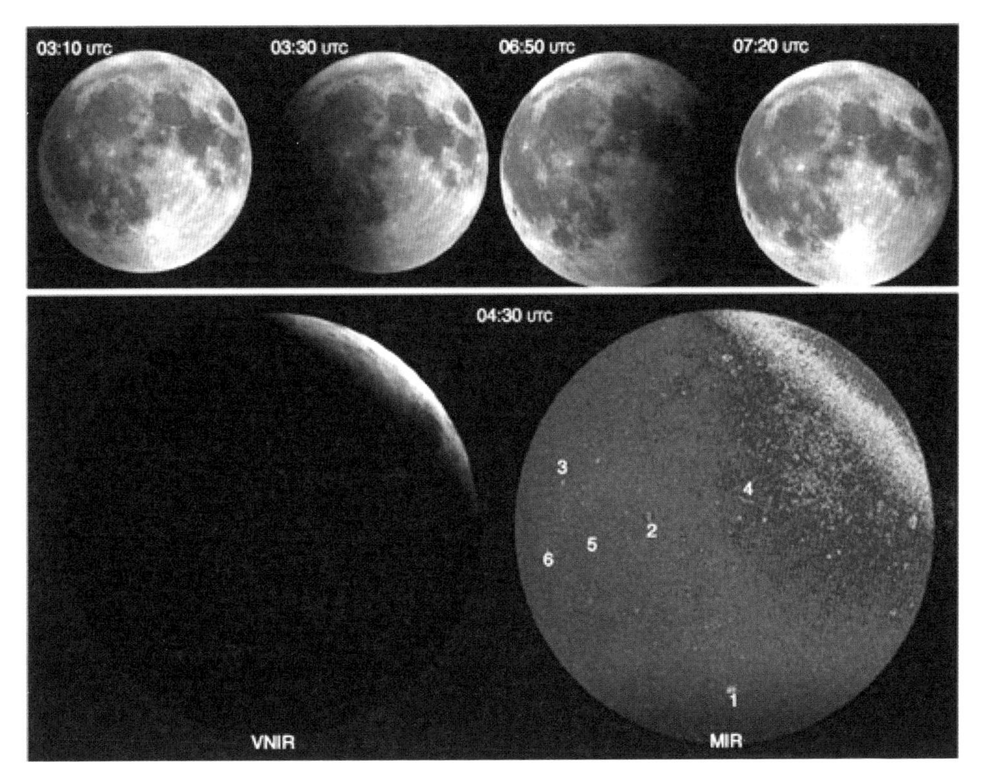

2019 年 1 月 21 日，紫金山天文台利用我国高分四号地球静止卫星多谱段观测到的月全食。

The total lunar eclipse observed by Purple Mountain Observatory using the Chinese Gaofen-4 Geo-Stationary Satellite with multiple spectral bands on 2019.1.21.

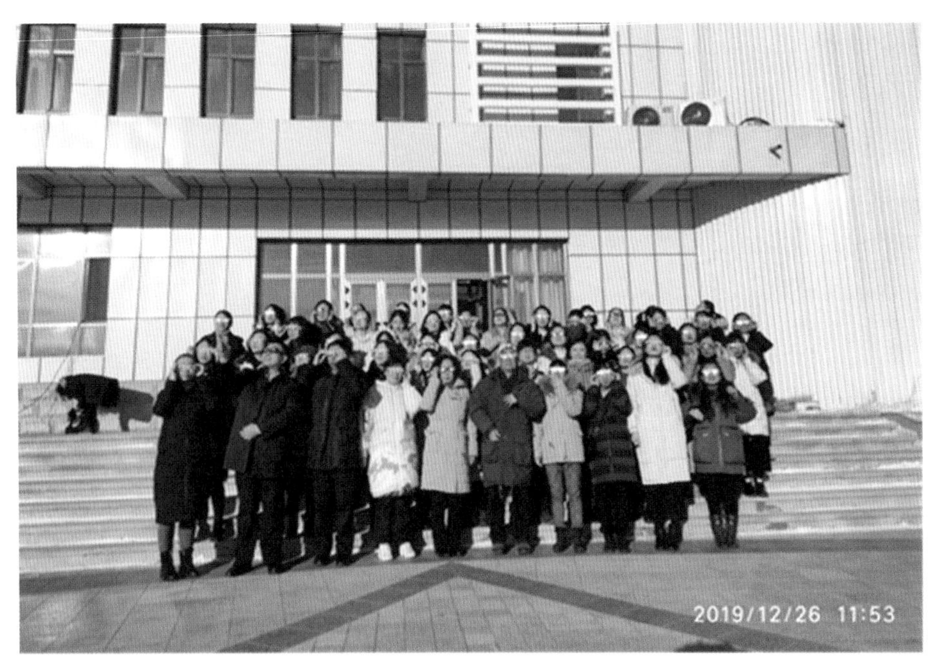

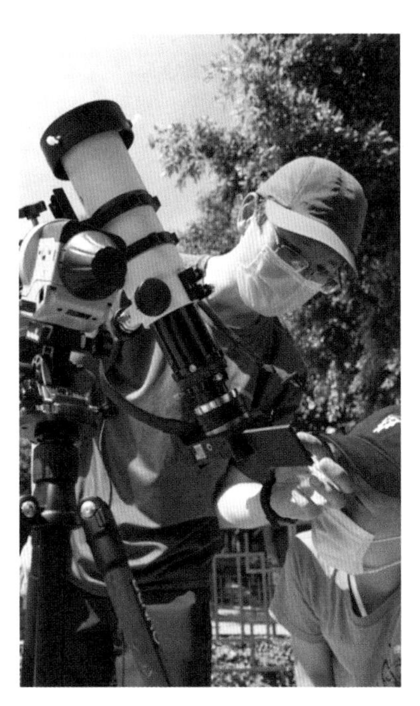

2019 年 12 月 26 日，第二届女天文工作者交流会参会代表在观测日偏食。

The participants of the second symposium for female astronomers observed the partial solar eclipse at the Sun Plaza, Nanshan Station on 2019.12.26.

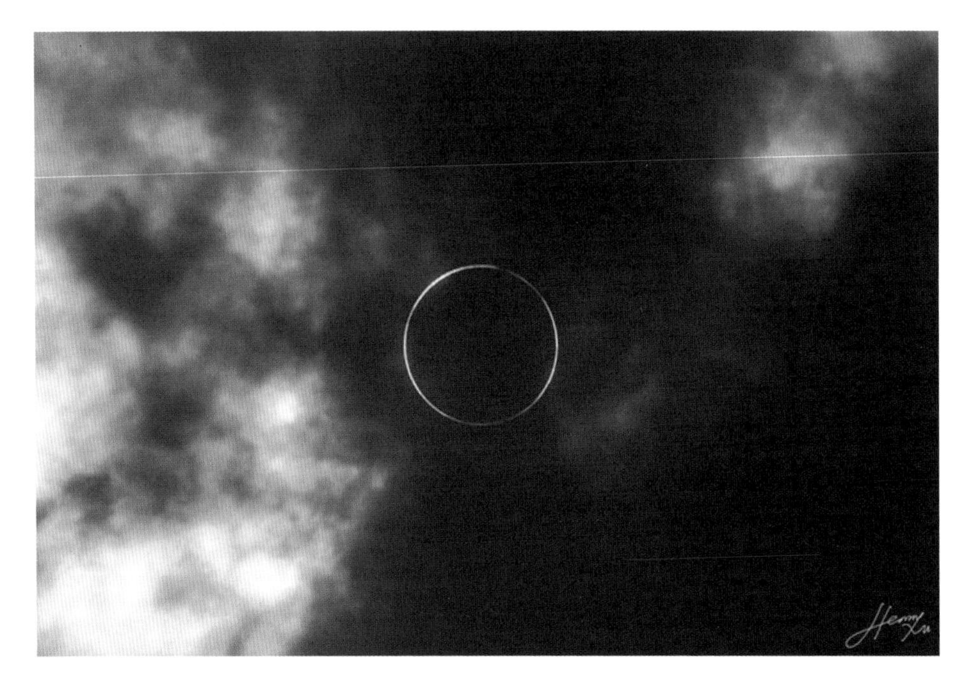

2020 年 6 月 21 日，我国南方出现日环食，全国各地十多万人赴厦门观测。

On 21 June 2020, an annular eclipse happened in southern China. Millions of people came to Amoy to observe.

五、学术活动 Academic Activities

2012 年美国 TMT 代表团访问南京天光所。

The TMT delegation of USA visited Nanjing Institute of Astronomical Optics and Technology in 2012.

上海交通大学天文系 2013 年 3 月主办了 BigBOSS［DESI（Dark Energy Spectroscopic Instrument）前身］合作会议。

In March 2013, Department of Astronomy of Shanghai Jiao Tong University organized the BigBOSS [predecessor of DESI (Dark Energy Spectroscopic Instrument)] collaboration meeting.

2013 年 9 月，第五届海峡两岸天文望远镜及仪器研讨会。

The 5th Cross-strait Astronomical Telescope and Instrument Symposium was successfully held in Xishuangbanna, Yunnan on 2013.9.3-6.

2013 年 10 月 17—18 日，“2013 年全国时间频率学术交流会”在陕西延安举行。

On October 17-18, 2013, “2013 National Time and Frequency Academic Conference” was held in Yan’an, Shaanxi.

2013 年 12 月，首届“南极星”青年科学家论坛。

The first term of “South Pole Star”forum of young scientists in 2013.12.

2014 年，国家天文台与乌兹别克斯坦天文台签署合作协议。

In 2014 the cooperation agreement between NAOC and Uzbekistan Observatory was Signed.

2014 年 7 月 17–18 日，JLRAT 实验室成员协助科技部在北京香山饭店召开以“中国射电天文学发展与平方公里阵列望远镜（SKA）”为主题的第 S21 次香山科学会议。

Members of the JLRAT laboratory assisted the Ministry of Science and Technology to hold the S21 Xiangshan Science Conference with the theme of “China’s Radio Astronomy Development and the Square Kilometre Array Telescope (SKA)” at the Xiangshan Hotel in Beijing on 2014.7.17-18.

第 22 届国际天文馆学会（IPS）大会（2014 年 6 月，北京）。

The 22nd International Planetarium Society Conference (June, 2014, Beijing).

2015 年 1 月 7 日，云南天文台与泰国国家天文研究所合作协议签字仪式在泰国首都曼谷举行。

The signing ceremony of the cooperation agreement between Yunnan Astronomical Observatory and Thailand National Astronomical Institute was held in the capital Bangkok of Thailand on 2015.1.7.

2015 年 8 月 30 日 –9 月 4 日，第六届海峡两岸天文望远镜与观测前沿技术研讨会在台湾花莲成功召开。

August 30 - September 4, 2015, the 6th Cross-strait Seminar on Advanced Technology of Astronomy Telescope and Observation was successfully held in Hualien, Taiwan.

2015 年 10 月 19 日 –21 日，中国天文学会 2015 年学术年会在北京大学召开。

The 2015 annual academic meeting of the Chinese Astronomical Society was held in Peking University from October 19 to October 21, 2015.

2016 年，国家天文台与西班牙加纳利大望远镜（GTC）签署合作协议。

In 2016 the cooperation betwenn NAOC and GTC project of Spain was signed.

2016 年 4 月 12–15 日，在南京主办了“第 27 届国际空间太赫兹技术研讨会（ISSTT）”，本次会议是 ISSTT 会议首次在中国召开。

The “27th International Symposium on Space Terahertz Technology (ISSTT)” was held in Nanjing on 2016.4.12-15. The ISSTT conference was first time held in China.

2016年6月7日，中美大学天文合作高峰论坛在北京大学举行。

The China-US Universities Astronomy Collaboration Summit, with the theme "Our Universe, a World without Boundaries", was held in Beijing on June 7, 2016.

2016年10月15日，“时间频率与卫星导航学术研讨暨庆祝国家授时中心成立50周年”在西安临潼国家授时中心召开。

On October 15, 2016, "Time, Frequency and Satellite Navigation Academic Seminar and Celebration of the 50th Anniversary of the Establishment of National Time Service Center" were held at National Time Service Center in Lintong, Xi'an.

2017 年，南非天文台代表团访问南京天光所。

In 2017 the delegation of South-Africa Observatory Visited Nanjing Institute of Astronomical Optics and Technology.

2017 年 5 月 4 日，北京大学科维理天文与天体物理研究所举办学术讨论会庆祝成立十周年。

The Kavli Institute for Astronomy and Astrophysics, Peking University (KIAA)'s 10th Anniversary Symposium was held at KIAA, Peking University on May 4, 2017.

2017 年 5 月 15−19 日，第一届中欧太阳物理会议在云南省昆明市成功举办。

The 1st China-Europe Symposium on Solar Physics was held successfully in Kunming, Yunnan Province on 2017.5.15-19.

2018 年 1 月，青年天文学者学术研讨会，云南昆明。

Workshop for Young Astronomers, Kunming, Yunnan, in January 2018.

2018 年 8 月，第七届“青年天文论坛”，新疆乌鲁木齐。

The 7th “Youth Astronomy Forum”, Urumqi, Xinjiang, August 2018.

2018 年 11 月 5 日，由中国卫星导航系统委员会主办，国家授时中心承办的联合国全球卫星导航系统国际委员会（ICG）第十三届大会在陕西西安开幕。相里斌副院长致辞。

On November 5, 2018, the 13th United Nations International Committee on Global Navigation Satellite Systems (ICG), hosted by the China Navigation Satellite System Committee and undertaken by National Time Service Center, opened in Xi’an, Shaanxi. CAS vice President Xiang Libin delivered a speech.

2018 年 12 月 10-14 日，第十届 PFS（PrimeFocus Spectrograph）合作会议。

The 10th PFS (PrimeFocus Spectrograph) collaboration meeting was held in December, 2018.

2019 年三十米望远镜（TMT）国际合作项目首光仪器团队成员访问南京天光所。

In 2019 the members of first-light instrumental term of 30-m telescope (TMT) visited Nanjing Institute of Astronomical Optics and Technology.

2019 年澳大利亚天文台（AAO）代表团来访南京天光所。

In 2019 the delegation of Australian Astronomical Observatory (AAO) visited Nanjing Institute of Astronomical Optics and Technology.

2019 年 6 月 3–5 日，2019 年全国 VLBI（甚长基线干涉测量技术）科学技术及应用研讨会在国家授时中心举行。

From June 3-5, 2019, the 2019 National VLBI (Very Long Baseline Interferometry) Science, Technology and Application Seminar was held at National Time Service Center.

2019 年 6 月 30 日 –7 月 5 日，国际天文联合会第 353 号学术讨论会，会议主题为“大型巡天观测时代的星系动力学”。

The IAU Symposium 353 was held in Shanghai on 2019.6.30-7.5.

2019 年 8 月，第八届“青年天文论坛”，山东青岛。

The 8th “Youth Astronomy Forum”, Qingdao, Shandong, August 2019.

2019 年 8 月 5–9 日，“Ia 型超新星前身星”国际学术会议在云南丽江举办。

The international symposium “Ia Supernova and Progenitor” was held in Lijiang, Yunnan, on 2019.8.5–9.

2019 年 9 月 2 日，第八届海峡两岸天文望远镜与观测前沿技术研讨会在台湾高雄召开。

On September 2 2019, the 8th Cross-strait Symposium on Telescope and Observation Frontier Technology was held in Kaohsiung, Taiwan.

2019 年 11 月 4-8 日，由上海交通大学、李政道研究所和上海天文台联合举办的国际会议“星系与宇宙学上海合作会议”。

The international conference on cosmology and galaxy formation was jointly organized by the Department of Astronomy at Shanghai Jiao Tong University, the Tsung-Dao Lee Institute, and Shanghai Astronomical Observatory on November, 4-8, 2019.

2019 年 12 月 16 日，第一届全国天文光子学学术研讨会在南京天文光学技术研究所举行。

The first National Symposium on Astronomical Photonics was held in Nanjing Institute of Astronomical Optics and Technology on December 16, 2019.

2020 年 10 月 12-14 日，“中国天文学会 2020 年学术年会”。本次年会采取线上线下相结合的形式举行，年会总注册人数超过 1000 人，成为历届天文学年会注册人数最多的一届。

“Chinese Astronomical Society 2020 Annual Academic Conference” from October 12-14, 2020. The Conference was held in a hybrid way (both online and offline) with more than 1,000 registrants, making it the largest number of registrants ever for an annual astronomy conference.

六、天文教育 Astronomical Education

2012年11月26日，举办厦门大学天文学系复办仪式。

On November 26, 2012, the Re-establishing Ceremony of the Department of Astronomy, Xiamen University was conducted.

2017 年 4 月 6 日，上海交通大学物理与天文学院召开天文系成立大会。

The establishment of the Department of Astronomy meeting was organized by Shanghai Jiao Tong University on April 6, 2017.

2017 年 11 月 26 日 –12 月 9 日，海上丝绸之路国际青年天文学者培训班在昆明成功举办。

The training class of international young astronomers for the Maritime Silk Road was held successfully in Kunming on 2017.11.26-12.9.

2019 年 4 月 21 日，清华大学天文系成立大会正式举行。

On April 21, 2019, the ceremony for establishment of the Department of Astronomy at Tsinghua University was held.

2019 年 10 月 14 日 -11 月 2 日，第 42 届国际青年天文学家培训学校在昆明由国际天文联合会（IAU）和中国科学院云南天文台联合主办。

The 42th international training school of young astronomers was held by IAU and Yunnan Astronomical Observatory in Kunming on 2019.10.14-11.2.

2019 年 12 月 7 日，中山大学物理与天文学院天文系在中山大学珠海校区复办揭牌仪式。

On December 7, 2019, the department of astronomy of the school of physics and astronomy of Sun Yat-sen University unveiling ceremony was held at the Zhuhai Campus of Sun Yat-sen University.

2020 年 1 月 6–10 日，由欧洲天文期刊《天文与天体物理》、EDP 出版社、中国科学院云南天文台联合主办的第 5 届青年天文工作者科技论文写作学校在中国科学院云南天文台凤凰山本部顺利举办。

The 5th writing school of scientific papers for young astronomers was organized by the European Journal *Astronomy and Astrophysics*, EDP Press, and Yunnan Astronomical Observatory in Kunming on 2020.1.6-10.

七、天文普及 Astronomical Popularization

2012 年 2 月 22 日，北京市人民政府天安门地区管理委员会向紫金山天文台赠送国旗，紫金山天文台天文历算研究人员按时为天安门管委会提供下一年度北京天安门地区每天日出日落的标准时刻。

On 2012.2.22, the Tiananmen District Management Committee of the Beijing Municipal People's Government presented the national flag to Purple Mountain Observatory, and the astronomical calendars researchers of Purple Mountain Observatory will provide the Tiananmen Square Management Committee with the standard sunrise and sunset times for the following year.

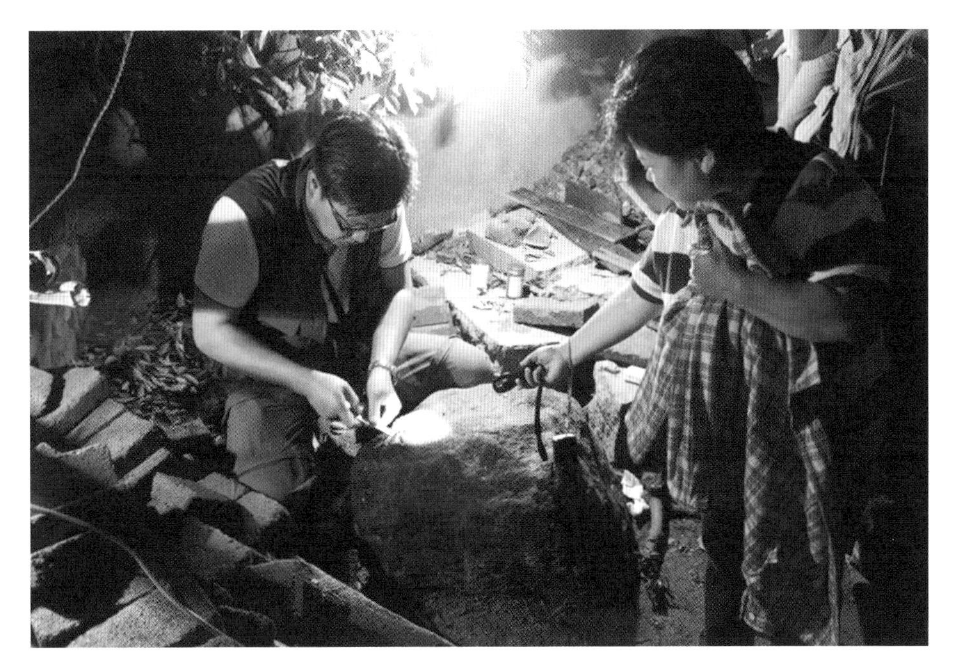

2012 年 5 月 20 日，海峡两岸陨石专家鉴定福建省福清东翁 585 公斤陨铁，此为我国私人藏家手上质量最大的陨铁标本。

On May 20, 2012 in Fuqing, meteorite experts from both sides of the Taiwan Straits had authenticated a 585kg siderite , which was the heaviest private-owned siderite specimen in China.

2012 年 10 月 26–27 日，“天文科普入寺院、医疗健康进藏区”活动在香格里拉成功举办。

The activity of “Astronomical Popularization into Temples, Medical Healthy into Tibet” was held in Shangri-La successfully in 2012.10.

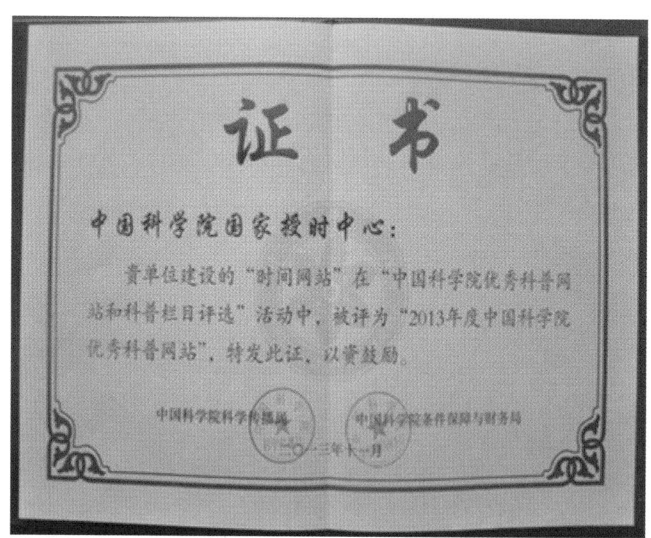

证 书

中国科学院国家授时中心：

贵单位建设的“时间网站”在“中国科学院优秀科普网站和科普栏目评选”活动中，被评为“2013年度中国科学院优秀科普网站”，特发此证，以资鼓励。

中国科学院科学传播局　中国科学院条件保障与财务局

二〇一三年十一月

2013 年国家授时中心“时间网站”被评为“2013 年度中国科学院优秀科普网站”。

The “time.ac.cn” of National Time Service Center was awarded “2013 Excellent Science Popularization Website of Chinese Academy of Sciences”.

2013 年 8 月，福建省天文学会与高雄市天文学会合作办理“闽台青少年天文夏令营”。图为台湾各地师生在中国天文学会创始人高鲁先生故里——福建省长乐县龙门村参访高家祠堂。

In August 2013, Fujian Astronomy Society has collaborated with Kaohsiung Astronomy Society in organizing Min-Tai Astronomy Youth Camp. The picture shows the scene of teachers and students from Taiwan in the hometown of Mr. Lu Gao, who was the founder of China Astronomy Society.

2013 年 8 月，闽台两省营员参访福州青少年活动中心天象馆。

The picture was the members of the youth camp visiting the planetarium in Fuzhou Youth Activity Center in August 2013.

斯洛文尼亚共和国总理米罗・采拉尔到访北京古观象台（2015 年 11 月，北京）。

Miro Cerar, the prime minister of the Republic of Slovenia, visited Beijing Ancient Observatory (November, 2015, Beijing).

2014 年 5 月青岛观象台被山东省科协、山东省财政厅命名为：四星级山东省科普教育基地。

Tsingtao Observatory was named by Shandong Provincial Science Association and Shandong Provincial Finance Department as: four-star Shandong Provincial Science Education Base in 2014.5.

澳大利亚总理马尔科姆・特恩布尔到访北京古观象台（2016 年 4 月，北京）。

Malcolm Turnbull, the prime minister of Australia, visited Beijing Ancient Observatory (April, 2016, Beijing).

国家授时中心 2017 年 5 月获"首批中国十大科技旅游基地"之一。

National Time Service Center was awarded one of the "First China's Top Ten Science and Technology Tourism Bases" in May 2017.

2017 年 8 月上旬福建省天文学会联合全台高校天文社联盟办理"闽台大学生天文交流营"， 图为两岸营员参观泉州一中天文台。

In early August 2017, Fujian Astronomical Society had collaborated with Astronomy Club Union of Universities in Taiwan in organizing Fujian-Taiwan Undergraduate Astronomy Communication Camp. The picture showed the scene of the members visiting Quanzhou No.1 High School Observatory.

2017 年 8 月中旬，闽台港澳四地天文同仁共聚一堂庆祝嘉义市天文协会成立三十周年。

图为莅会嘉宾参与路边天文活动。

In middle August 2017, colleagues from Fujian, Taiwan, Hong Kong and Macao had grouped together to celebrate 30th Anniversary of Chiayi Amateur Astronomers Association. The picture showed the scene of the guests participating in roadside astronomy activities.

2019 年 6 月，由国家授时中心策划的“中国古代导航展”亮相联合国维也纳中心。

In June 2019, “Ancient Chinese Navigation Exhibition” planned by National Time Service Center was unveiled at the United Nations Vienna Center.

北京天文馆与埃及亚历山大图书馆天文馆在“一带一路”科普场馆发展国际研讨会上签订合作谅解备忘录（2019 年 10 月，北京）。

Beijing Planetarium and the Planetarium Science Center Bibliotheca Alexandrina signed a MOU on the International Symposium on the Development of Natural Science Museums under the Belt and Road Initiative (October, 2019, Beijing).

中国大陆地区 IAU100 活动启动仪式（2019 年 12 月，北京）。

The opening ceremony of the activities of IAU100 in the Chinese mainland was held in Beijing, 2019.12.

2021 年 7 月，上海天文馆新馆教育空间“天文数字实验室”开放。

In July 2021, “Astronomical digital laboratory” of Shanghai Planetarium was opened.

2021 年 8 月福建省天文学会建成永泰县盖洋乡“乌石庄园”星空研学观测基地。

In August 2021, Fujian Astronomical Society had founded Wushi Manor, an observation base in Gaiyang, Yongtai.

2021 年南京天文光学技术研究所顺利完成“科技冬奥”冰状雪赛道测试工作。

In 2021 Nanjing Institute of Astronomical Optics and Technology had completed the test of glacial snow track for “Science and Technology of Winter Olympics ”.

编后语

值此中国天文学会成立百年纪念之际，中国天文学会理事会决定成立《中国天文学在前进》第五册纪念文集编委会，本着对学会负责，对历史负责和对天文事业负责的原则着手征集稿件、收集资料、翻译校对、修改润色、图片选取等系列准备工作。编委会与学会会员单位及各专业委员会和工作委员会通力合作，并得到南京大学出版社大力支持，按时完成了纪念文集的出版。

第五册纪念文集真实地记录了中国天文学近十年所取得的重要成果和事件，在此谨以此书献给在天文学各领域辛勤工作的同仁，同时献给始终关注和支持中国天文学发展的领导、专家和社会各界的朋友，以及广大天文爱好者。

对参与本书撰稿、翻译、校对和图片资料提供的各单位，专业委员会和工作委员会，省市(及地方)天文学会和所有人员表示衷心感谢。对纪念文集中可能出现的错误或不当之处，恳请读者指正。

《中国天文学在前进》

第五册纪念文集编委会

2022年11月7日

Words of the Editors

During the time for the centennial anniversary of Chinese Astronomical Society (CAS), the council of CAS has decided to establish the board of editors for fifth volume of commemoration 《Chinese Astronomy on the Match》. In line with the principal to be responsible for CAS, the history, and the Chinese astronomical careers, it's started to prepare a series of the collection of manuscripts and material, translation and correction, selection of pictures and so on.

Under the effective cooperation with the units, professional and working committees of CAS, and strongly supported by the Press of Nanjing University, the commemorative has been published on time.

The fifth volume of Commemoration has recorded exactly the important achievements and events of Chinese astronomy in recent ten years, which is dedicated to Chinese colleague who hard working in various fields of astronomy, and the leaders, experts, and friends of all sectors of society, as well as broad amateurs of astronomy who always pay attention to and support Chinese astronomy.

It's sincerely grateful to various units, professional and working committees, the local members of CAS, and all the people who contribute to the manuscripts, translations, and corrections, and provide pictures and relevant material. Welcome all the readers to criticize and correct the possible mistakes or improper places.

The editorial board of fifth volume of Commemoration
《Chinese Astronomy on the Match》
November 7, 2022